51单片机
项目教程

C语言版 | 赠单片机开发板

—— 吴险峰 主编 ——
但唐仁 刘德新 曾路 副主编

人民邮电出版社
北　京

图书在版编目（CIP）数据

51单片机项目教程：C语言版：赠单片机开发板 / 吴险峰主编. -- 北京：人民邮电出版社，2016.8（2021.9重印）
ISBN 978-7-115-42540-9

Ⅰ. ①5… Ⅱ. ①吴… Ⅲ. ①单片微型计算机—高等学校—教材 Ⅳ. ①TP368.1

中国版本图书馆CIP数据核字(2016)第139479号

内 容 提 要

本书以 51 系列单片机为载体，采用项目化教学方式，由浅入深地安排实训内容，将知识点和相关实训内容结合，突出对读者动手能力的培养。项目内容包含了基础和拓展两大部分。基础部分包括单片机开发环境构建、流水灯、蜂鸣器、LED 显示器、按键输入、中断、定时器和串口通信等内容，免费赠送的开发板可设计实验完全覆盖这些内容；拓展部分则是精选出的具有代表性的真实项目，体现了时代性的创客特色，包括红外遥控、声音传感器、温度传感器、倾斜开关、超声传感器、人体红外传感器、火焰传感器和无线模块等内容。同时，本书配套的网站提供了更多的实训项目资源，进一步提高读者实战技能，体现"互联网+"特色。

本书适合单片机初学者使用，也适合作为高职高专院校单片机技术课程教材。

◆ 主　编　吴险峰
　　副 主 编　但唐仁　刘德新　曾　路
　　责任编辑　刘盛平
　　执行编辑　左仲海
　　责任印制　焦志炜

◆ 人民邮电出版社出版发行　北京市丰台区成寿寺路11号
　　邮编　100164　电子邮件　315@ptpress.com.cn
　　网址　http://www.ptpress.com.cn
　　北京市艺辉印刷有限公司印刷

◆ 开本：787×1092　1/16
　　印张：16　　　　　　　　　2016年8月第1版
　　字数：454千字　　　　　　2021年9月北京第12次印刷

定价：99.80元（含开发板）

读者服务热线：(010)81055256　印装质量热线：(010)81055316
反盗版热线：(010)81055315

"互联网+"创新教材编委会委员（排名不分先后）

吴险峰	但唐仁	李华忠	郑洪英	盛建强
曾 路	张宗平	刘德新	余柏林	陈明芳
李 力	赖友源	孙 波	侯继红	刘国成
江武志	刘 洋	化雪荟	许政博	李永祥

序

　　教材是进行教学的基本工具，也是深化教育教学改革、全面推进素质教育、培养创新人才的重要保证。教育部也把教材建设作为衡量高职高专院校深化教育教学改革的重要指标，作为检验各高职院校人才培养工作的质量与力度。

　　针对"互联网+"国家战略，本教材编委会对嵌入式专业类课程进行了大胆的尝试，与产业界和兄弟院校联合探索课程发展的"互联网+"模式——购教材配开发板，厂家不通过开发板盈利，而是靠后期的课程特色资源、开发板配件和增值服务等来获得收益，并依照该模式创新的推出中国第一套直接送开发板的系列教材，实现嵌入式专业的"四化"目标。

1. 教与练一体化

　　本教材内容和开发板严格匹配，实现课程知识目标和技能训练紧密结合，理论教学和实战演练一体化操作，更有助于学生的技能掌握。

2. 授课云端化

　　传统嵌入式类课程，往往需要专门的实训箱甚至是定制的实训平台，才能进行实训操作。本教材自带开发板，不需要准备专门的机房，也不担心对实训设备的维护保养，甚至学生可以通过网络课堂进行远程学习和实训，降低了成本，提高了学习效率。

3. 学生创客化

　　很多创客都来自嵌入式专业，但若只是通过学校实验箱来被动学习，没有属于自己的开发板，是不可能成为一个创客的。现在教材配送的开发板归属学生所有，学生有思路就可以马上实践，从而诞生更多的创客灵感。

4. 认证标准化

　　相对于标准化的计算机等级水平考试和软件类考试来说，嵌入式类课程的开发板没有统一的标准，不方便组织统一的职业技能实操考试，只能进行理论考试认证。而通过该模式，每个采用相关教材的学生都是采用同一开发板进行技能训练，有助于职业技能认证的标准化。

　　创新肯定会碰到各种问题，希望本系列教材在实践中不断修订完善，促进嵌入式技术教育蓬勃发展，同时对软硬件结合的其他专业也能有重要的启示。

高林
2016 年 4 月

前言 FOREWORD

近来职业院校的单片机课程改革力度很大,大部分教材注重工学结合、校企合作,通过项目教学模式来加强实践和职业技能培养。但如何将教材和学习开发板相结合,一直是一个难点。由于各校的开发板不统一,没有标准化。嵌入式专业的教材要么是和学校实训室的开发板捆绑,要么是通过纯软件完成虚拟实训内容。前者虽然针对性强,但需要配备专门实验条件,而且教材发行量有限。后者虽便于教学,但是对于学生而言,离实际技能培养还是有相当差距。如何让教材发挥真正的作用,是每个专业老师都非常关心的话题。

本教材最大特色就是结合"互联网+"国家战略对教材进行大胆的探索。创新"互联网+"教材模式,打造全新的单片机学习生态链,提高学习效率,降低学习成本。

具体来说,教材分成基础部分和拓展部分。对于教学必须完成的基础部分内容,教材配套提供试验用开发板。对于拓展部门的内容,读者需要用到的其他配件和模块,需要另行购买。

全书包括单片机开发环境构建、流水灯、蜂鸣器、LED 显示器、按键输入、中断、定时器和串口通信等内容,通过配套的开发板完全覆盖,教材的内容结构如下。

基 础 部 分		拓 展 部 分	
项目 1	构建单片机开发环境	项目 13	LCD 显示器静态显示字符
项目 2	了解开发板	项目 14	红外遥控
项目 3	点亮一个 LED	项目 15	简易计算器(LED 显示器显示)
项目 4	流水灯实验	项目 16	音乐喷泉
项目 5	蜂鸣器实验	项目 17	防盗报警器
项目 6	LED 显示器静态显示	项目 18	8×8 点阵显示"爱心"
项目 7	LED 显示器动态显示	项目 19	温度计显示
项目 8	独立键盘输入	项目 20	测距显示
项目 9	单片机中断系统	项目 21	步进电机控制
项目 10	单片机定时器	项目 22	A/D-D/A 模块
项目 11	串口通信	项目 23	火焰报警器
项目 12	综合项目:秒表	项目 24	人体红外感应灯
		项目 25	无线模块
		项目 26	智能风扇系统(综合实验)

本教材保持了最大的灵活性,可以根据课程的不同学时,设置不同的项目。企业可以通过互联网发布最新的项目资源,同时为感兴趣的同学提供项目和技术支持。这种全新的"互联网+"的校企合作,相信也会对整个行业产生重要的引领作用。另外,标准化的开发板也有利于组织标准化的技能竞赛和认证考试。相对于软件认证考试来说,硬件认证考试一直是整个行业的痛点。本教材也在这方面进行大胆的尝试。

FOREWORD

  本教材的诞生凝聚了深圳信息职业技术学院软件学院嵌入式专业老师的多年心血。吴险峰对本教材内容的编写与项目设计进行了总体策划,对全教材进行了统稿和初审,但唐仁、刘德新、曾路、李华忠、郑洪英和盛建强等参与了书稿整理和在线教育网站的制作。本教材在编写的过程中还得到了深圳信息职业技术学院张宗平和余柏林、广东科学技术职业学院陈明芳和李力、广东工程职业技术学院赖友源、广东青年职业学院孙波、广州科技贸易职业学院侯继红、广州铁路职业技术学院刘国成、中山职业技术学院江武志、东莞职业技术学院刘洋、佛山职业技术学院化雪荟等各位同人的大力支持,在此一并表示衷心的感谢!

  本教材同时得到了相关企业的大力支持。深圳市亚博智能科技有限公司不仅为教材配赠了开发板,还专门为此项目录制了视频,同时也提供相关增值服务。珠海因尔科技有限公司为本项目提供进一步发展的平台,将此创新理念进一步拓展到嵌入式和物联网相关领域。深圳市华谕电子商务有限公司为教材配套网站提供技术支持。在此一并表示衷心的感谢!

  该教材电子课件、电子教案、全套视频、实例程序和辅助工具等教学素材,可登录 www.ryjiaoyu.com 免费下载。试题库、认证考试和竞赛库等其他教学资源也在逐步完善,希望为大家提供一整套的完整教学资源。后期还会开通微信服务号,开发相应的移动应用 App,构建完整的单片机生态链。

编　者
2016 年 4 月

目录

CONTENTS

项目 1　构建单片机开发环境　1
  1.1　项目分析　2
  1.2　技术准备　2
    1.2.1　单片机系统简介　2
    1.2.2　安装 USB 驱动程序　2
    1.2.3　单片机程序烧录方法和相关工具　3
    1.2.4　使用单片机开发软件 Keil4　5

项目 2　了解开发板　8
  2.1　项目分析　9
  2.2　技术准备　9
    2.2.1　亚博 BST-M51 模块电路图　9
    2.2.2　亚博科技 BST-M51 学习板功能模块图　10

项目 3　点亮一个 LED　11
  3.1　项目分析　12
  3.2　技术准备　12
    3.2.1　理论知识　12
    3.2.2　程序基础　12
  3.3　项目实施　26

项目 4　流水灯实验　29
  4.1　项目分析　30
  4.2　技术准备　30
    4.2.1　流水灯硬件实物　30
    4.2.2　流水灯原理图　30
    4.2.3　流水灯实验理论知识　31
    4.2.4　定时器的结构　33
    4.2.5　定时器的 TMOD 和 TCON 寄存器　33
    4.2.6　定时器工作方式　34
    4.2.7　定时器编程步骤　34
  4.3　项目实施　35

项目 5　蜂鸣器实验　38
  5.1　项目分析　39
  5.2　技术准备　39
    5.2.1　蜂鸣器硬件实物　39
    5.2.2　蜂鸣器实验相关电路　39
    5.2.3　串行接口的结构　40
    5.2.4　串口的 4 种工作方式和波特率　41
    5.2.5　C51 串口编程　43
  5.3　项目实施　43
    5.3.1　简单蜂鸣器发声实验　43
    5.3.2　给前面任务的流水灯加入报警效果　44

项目 6　LED 显示器静态显示　46
  6.1　项目分析　47
  6.2　技术准备　47
    6.2.1　LED 显示器静态显示简介　47
    6.2.2　LED 显示器可显示内容和特点　47
    6.2.3　LED 显示器的结构与原理　47
    6.2.4　LED 显示 2 种接法　48
  6.3　项目实施　49

项目 7　LED 显示器动态显示　51
  7.1　项目分析　52
  7.2　技术准备　52
    7.2.1　LED 显示器动态显示与扫描原理　52
    7.2.2　4 位 LED 显示器的动态和静态显示连接方式图　52
    7.2.3　编程实验理论准备　52
  7.3　项目实施　55

项目 8　独立键盘输入　58
  8.1　项目分析　59
  8.2　技术准备　59

CONTENTS

　　8.2.1　独立键盘输入理论知识　59
　　8.2.2　硬件模块工作原理　59
　　8.2.3　认识轻触开关按键　60
　　8.2.4　键盘的分类　61
8.3　项目实施　61

项目 9　单片机中断系统　63
9.1　项目分析　64
9.2　技术准备　64
　　9.2.1　单片机中断系基本概念　64
　　9.2.2　中断传送方式及其特点　64
　　9.2.3　80C51 中断系统　65
　　9.2.4　中断服务函数　68
9.3　项目实施　69
　　9.3.1　外部中断低电平触发　69
　　9.3.2　外部中断下降沿触发　70

项目 10　单片机定时器　71
10.1　项目分析　72
10.2　技术准备　72
　　10.2.1　单片机定时基础　72
　　10.2.2　定时/计数器 T0 的工作原理　72
　　10.2.3　定时/计数器的结构及工作原理　72
　　10.2.4　定时/计数器的工作方式　75
10.3　项目实施　77

项目 11　串口通信　79
11.1　项目分析　80
11.2　技术准备　80
　　11.2.1　串口通信理论知识　80
　　11.2.2　计算机通信的分类　80
　　11.2.3　串行通信与并行通信　80
　　11.2.4　信号的调制与解调　82
　　11.2.5　串行通信的错误校验　82
　　11.2.6　传输速率及其与传输距离的关系　83
　　11.2.7　串口结构　83
　　11.2.8　串行通信的数据结构　83

11.3　项目实施　85

项目 12　综合实验：秒表　87
12.1　项目分析　88
12.2　技术准备　88
12.3　项目实施　88

项目 13　LCD 显示器静态显示字符　93
13.1　项目分析　94
13.2　技术准备　94
　　13.2.1　LCD 1602 介绍　94
　　13.2.2　LCD1602 的驱动操作　95
　　13.2.3　LCD 1602 的指令码　96
　　13.2.4　RAM 地址映射图　97
13.3　项目实施　97

项目 14　红外遥控　102
14.1　项目分析　103
14.2　技术准备　103
　　14.2.1　红外线　103
　　14.2.2　NEC 协议　108
　　14.2.3　红外遥控器键码值　109
14.3　项目实施　109
　　14.3.1　红外线发送　109
　　14.3.2　红外线接收　113

项目 15　简易计算器（LED 显示器显示）　117
15.1　项目分析　118
15.2　技术准备　118
15.3　项目实施　118
15.4　技术拓展　120

项目 16　音乐喷泉　122
16.1　项目分析　123
16.2　技术准备　123
　　16.2.1　模块原理图　123
　　16.2.2　模块接口说明　124
16.3　项目实施　124

CONTENTS

16.4 技术拓展 125

项目 17　防盗报警器　127
17.1 项目分析 128
17.2 项目准备 128
17.3 项目实施 129

项目 18　8×8 点阵显示"爱心"　131
18.1 项目分析 132
18.2 技术准备 132
 18.2.1 8×8 点阵介绍 132
 18.2.2 MAX7219 介绍 132
 18.2.3 MAX7219 引脚说明 132
 18.2.4 串行数据格式 133
 18.2.5 可寻址的数据寄存器和控制寄存器 133
18.3 项目实施 134
18.4 技术拓展 136

项目 19　温度计显示　138
19.1 项目分析 139
19.2 技术准备 139
 19.2.1 DS18B20 单线总线的工作方式 139
 19.2.2 DS18B20 的操作步骤 142
19.3 项目实施 145
19.4 技术拓展 148

项目 20　测距显示　150
20.1 项目分析 151
20.2 技术准备 151
 20.2.1 HC-SR04 超声波测距模块 151
 20.2.2 超声波测距原理 152
20.3 项目实施 153
20.4 技术拓展 158
 20.4.1 超声波测距（LED 显示器显示改 I/O 端口） 158
 20.4.2 超声波测距 LCD1602 显示 159

项目 21　步进电机控制　161
21.1 项目分析 162
21.2 技术准备 162
 21.2.1 步进电机简介 162
 21.2.2 步进电机转动原理及内部结构 162
 21.2.3 ULN2003 163
21.3 项目实施 163
 21.3.1 单－双八拍 163
 21.3.2 加、减速 166
 21.3.3 双四拍 168
21.4 技术拓展 170
 21.4.1 正、反转 170
 21.4.2 速度调节 171
 21.4.3 自制秒表 174

项目 22　A/D-D/A 模块　177
22.1 项目分析 178
22.2 技术准备 178
 22.2.1 PCF8591 介绍 178
 22.2.2 PCF8591 的器件地址与控制寄存器 179
 22.2.3 I^2C 总线的数据传送 179
22.3 项目实施 184
22.4 技术拓展 190
 22.4.1 D/A 输出模块 190
 22.4.2 A/D 模块（LCD1602 显示） 190

项目 23　火焰报警器　193
23.1 项目分析 194
23.2 技术准备 194
 23.2.1 光、热敏电阻拓展接线原理 194
 23.2.2 火焰传感器介绍 194
23.3 项目实施 194
23.4 技术拓展 200
 23.4.1 热感灯 200
 23.4.2 火焰传感器报警 201

CONTENTS

项目 24　人体红外感应灯　　202
　24.1　项目分析　　203
　24.2　技术准备　　203
　　24.2.1　基本概念及参数　　203
　　24.2.2　功能特点　　203
　　24.2.3　使用说明　　204
　24.3　项目实施　　205
　24.4　技术拓展　　210

项目 25　无线模块　　211
　25.1　项目分析　　212
　25.2　技术准备　　212
　　25.2.1　NRF24L01 简介　　212
　　25.2.2　模块外接引脚　　212
　　25.2.3　SPI　　212
　　25.2.4　工作模式　　214
　　25.2.5　数据通道　　215
　25.3　项目实施　　216
　25.4　技术拓展　　224

项目 26　智能风扇系统（综合实验）　　225
　26.1　项目分析　　226
　26.2　技术准备　　226
　26.3　项目实施　　226

附件　亚博 BST-M51 主要模块电路图　　240

参考文献　　244

Chapter 1

项目1
构建单片机开发环境

项目目标

- 通过构造单片机开发环境，了解单片机开发系统结构和流程。

建议学时

- 2学时。

知识要点

- 单片机系统开发流程。
- 编程工具Keil C51。
- 程序烧录。

技能掌握

- 安装USB驱动；利用Keil C开发环境编辑、编译、调试C51程序的过程；掌握实用程序烧录方法及相关工具。

1.1 项目分析

学习单片机之前，必须要掌握构建单片机开发环境的方法。本项目详细讲解 USB 驱动程序的安装方法，如何安装使用开发软件 Keil4，如何烧录程序和使用相关工具。

1.2 技术准备

1.2.1 单片机系统简介

单片机（Microcontrollers）是一种集成电路芯片，它是采用超大规模集成电路技术把具有数据处理能力的中央处理器 CPU、随机存储器 RAM、只读存储器 ROM、多种 I/O 端口和中断系统、定时器/计数器（可能还包括显示驱动电路、脉宽调制电路、模拟多路转换器、A/D 转换器）等集成到一块硅片上构成的一个微型计算机系统，并在工业控制领域广泛应用。从 20 世纪 80 年代起，单片机已由当时的 4 位、8 位单片机，发展到现在的 32 位的主频超过 300M 的高速单片机。

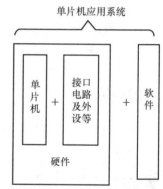

图 1-1 单片机应用系统结构

在单片机应用系统开发过程中，单片机是整个设计的核心。如图 1-1 所示，单片机应用系统由硬件和软件组成。硬件是应用系统的基础，软件在硬件的基础上对其资源进行合理调配和使用，从而完成应用系统所要求的任务，二者相互依赖，缺一不可。

1.2.2 安装 USB 驱动程序

确保计算机联网。USB 口插入开发板后，计算机会自动识别设备，自动联网，安装驱动设备。若计算机未自动安装驱动，打开设备管理器，找到插入开发板后弹出的新设备（未安装好驱动的情况下在"其他设备"中，带有黄色感叹号或问号），右键单击该设备，选择"更新驱动程序软件"→"自动搜索更新的驱动程序软件"即可安装驱动，更新驱动程序软件步骤如图 1-2～图 1-5 所示。

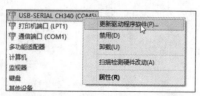

图 1-2 安装驱动程序软件步骤一

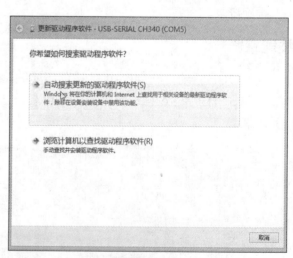

图 1-3 安装驱动程序软件步骤二

项目 1 构建单片机开发环境

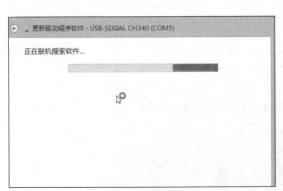

图 1-4 安装驱动程序软件步骤三

图 1-5 安装驱动程序软件步骤四

1.2.3 单片机程序烧录方法和相关工具

STC 单片机的程序烧录方法如下。

第一步,硬件连接。

将配套的 USB 电源线与串行端口线连上,在锁紧座上放入 STC 单片机,其他的硬件连接保持原状态,硬件连接完毕,打开电源。

第二步,软件操作。

(1)打开软件,出现如图 1-6 所示界面。

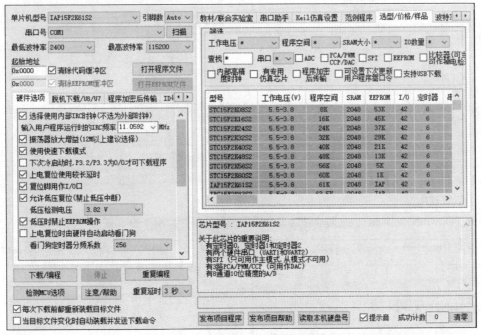

图 1-6 STC-ISP(V6.85)界面

(2)需要设置的几个参数,如图 1-7 所示。

参数设置在软件中都有详细提示,这里只是再次强调一下。

第三步,下载程序。

(1)导入源程序,可查看源代码,如图 1-8 所示。

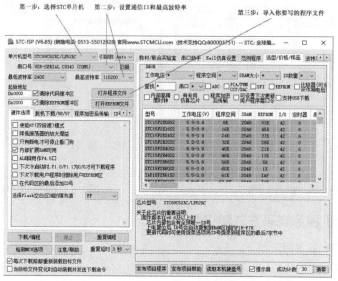

图1-7　STC-ISP（V6.85）设置参数

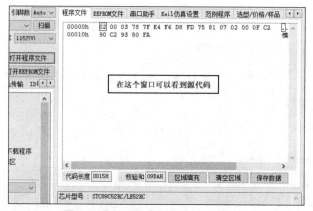

图1-8　STC-ISP（V6.85）查看代码窗口

（2）关闭单片机开发板套件电源为下载做准备。

（3）下载程序。

单击"Download/下载"，这时软件信息会提示操作过程，如图1-9和图1-10所示。

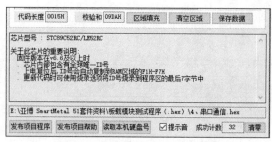

图1-9　STC-ISP（V6.85）下载过程

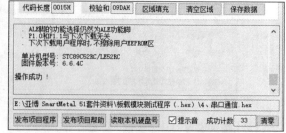

图1-10　STC-ISP（V6.85）烧录操作成功

（4）当提示窗口的内容为给MCU（微控制单元，又称单片机）通电时，打开单片机开发板套件的电源，程序将会自动完成写入。程序写入后，在本书附带的单片机开发板套件中，会立即演示出程序的效果。

1.2.4 使用单片机开发软件 Keil4

1. 编程工具 Keil C51

Keil Software 公司推出的 Keil μVision 是一款基于 Windows 的软件平台。它是一种用于 51 系列单片机的集成开发环境（Intergrated Development Enviroment，IDE）。目前的 μVision4 版本还可以支持 ARM 编程。μVision 提供了功能强大的编辑器和调试器，编辑器可以像一般的文本编辑器一样对源代码进行编辑，调试器使用户能够快速地检查和修改程序。用户还可以选中变量和存储器来观察其值，并可在双层窗口中显示，还可对其进行适当的调整。

Keil 同时支持 C 语言和汇编语言编程，本书只针对 C 语言进行讲解。Keil C51 编译器在遵循 ANSI C 标准的同时，为 51 单片机进行了特别的设计和扩展，可为用户提供在应用中需要的所有资源。

Keil C51 的库函数含有 100 多种功能，其中大多数是可载入的。函数库支持所有的 ANSI C 的程序。库函数中的程序还为硬件提供特殊指令，例如 nop、testbit、rol、ror 等，方便了应用程序的开发。

2. 用 Keil C51 编写程序

编程语言都需要开发环境，这样才能完成程序的编写、调试和编译。C51 程序开发是在 Keil μVision 开发环境下进行的。下面首先介绍该开发环境 Keil μVision4 的使用。该软件启动界面如图 1-11 所示。

（1）单击 "Project" → "New μVision Project" 新建一个工程，如图 1-12 所示。

图 1-11 Keil μVision4 软件启动界面

图 1-12 Keil μVision4 新建工程

（2）在对话框中，给这个工程取名 test1 后，保存，不需要填写后缀，如图 1-13 所示。注意，默认的工程后缀与 uVision3 及 uVision2 版本不同，为 uvporj。

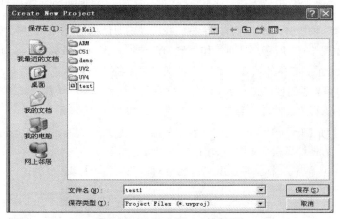

图 1-13 Keil μVision4 新建工程

（3）在弹出的对话框中，CPU 类型下选择实际使用的单片机型号。本例找到并选中"Atmel"下的 AT89S51，如图 1-14 所示。

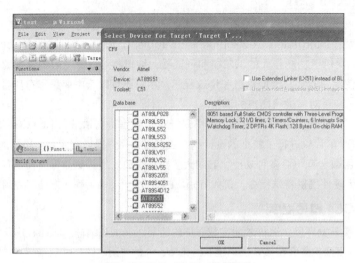

图 1-14　Keil μVision4 选择型号

（4）在以上工程创建完毕后，开始建立一个源程序文本，如图 1-15 所示。
（5）在源程序编辑区写入完整的 C 程序，如图 1-16 所示。

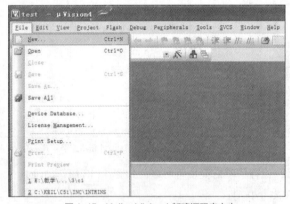　　　　　

图 1-15　Keil μVision4 新建源程序文本　　　　图 1-16　Keil μVision4 源程序编辑

（6）单击"保存"快捷键，弹出保存对话框，在文件名对话框里输入源程序文件名，此示例输入的文件名为"test"，如图 1-17 所示。注意，如果使用汇编语言，源程序文件名必须使用".asm"后缀，如本示例为"test.asm"；如果使用 C 语言，则源程序文件名必须使用".c"后缀，如本示例为"test.c"，然后保存。此时可以看到程序文本字体颜色已发生了变化，表明编译器生效。

（7）把刚创建的源程序文件加入到工程项目文件中，如图 1-18 所示。

（8）单击目标选项（Target Options）按钮"　"，先选择"Target（目标）"标签，设置晶振，一般设置为 12MHz，方便计算指令时间，如图 1-19 所示。

（9）在"Output（输出）"标签栏选中"Create HEX File"，使编译器的输出为单片机需要的 HEX 文件，如图 1-20 所示。

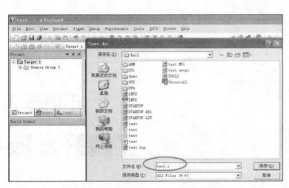
图 1-17 Keil μVision4 保存文件

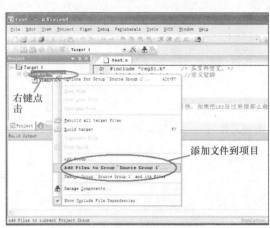
图 1-18 Keil μVision4 添加项目

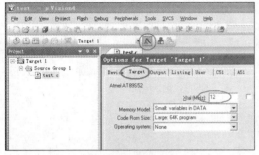
图 1-19 Keil μVision4 生成 HEX 文件

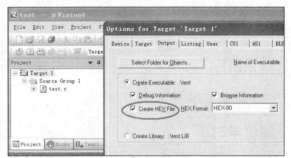
图 1-20 Keil μVision4 创建 HEX 文件

（10）保存后，单击重新建造（Rebuild）按钮" "进行编译，"Built Output（重建输出）"窗口在文件没有错误的情况下，提示有 hex 文件输出，如图 1-21 所示。

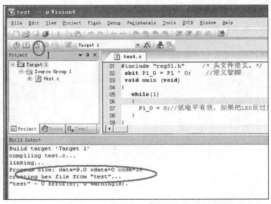
图 1-21 Keil μVision4 生成 HEX 文件

C51 程序开发与 Windows 中运行的项目工程的开发有所不同，在 Windows 中，一般程序的编译结果是后缀名为".exe"的可执行文件，该文件在 Windows 系统中能直接运行，而单片机 C51 程序的开发属于嵌入式开发，遵循主流的交叉编译模式，即在宿主机（运行 Keil μVision4 的 PC 机）上开发编译，在目标机（51 系列单片机）上运行。在宿主机上的编译结果为 HEX 文件，要经过编程器烧录到单片机的程序区（Flash ROM）才能执行。当然也可以通过 EDA 仿真软件来执行。

Chapter 2

项目2 了解开发板

项目目标
- 通过课堂讲解了解本开发板。

建议学时
- 2学时。

知识要点
- BST-M51电路图。
- BST-M51学习板功能模块图。

技能掌握
- 熟知BST-M51开发板。

2.1 项目分析

使用开发板之前，必须熟悉各个模块的电路图及功能。

2.2 技术准备

2.2.1 亚博 BST-M51 模块电路图

1. CH340 烧录模块电路

CH340 烧录模块电路如图 2-1 所示，图中 VCC-5V 为取自 USB 的 5V 电源；VCC 为模块供电点；TXD 为串行数据输出；RXD 为行数据输入；GND 为模块接地。

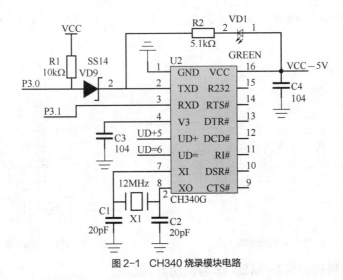

图 2-1 CH340 烧录模块电路

2. 4 位数码显示电路

4 位数码显示电路如图 2-2 所示，每位由 a～g7 段发光二极管（LED）组成。加正电压的发光，加零电压的不能发光，不同亮暗的组合就能显现不同的数字。

3. WIFI 模块电路

WIFI 模块电路如图 2-3 所示。该模块由单 5V 或 3.3V 电源供电；工作温度范围为 -45℃～+85℃；模块尺寸为 32mm×20mm×4.5mm。

4. 通用红外遥控系统模块电路

通用红外遥控系统模块电路由发射和接收两大部分组成，如图 2-4 和图 2-5 所示。应用编 / 解码专用集成电路芯片进行控制操作。发射部分包括键盘、编码调制、LED 红外发送器；接收部分包括光 / 电转换放大器、解调、解码电路。

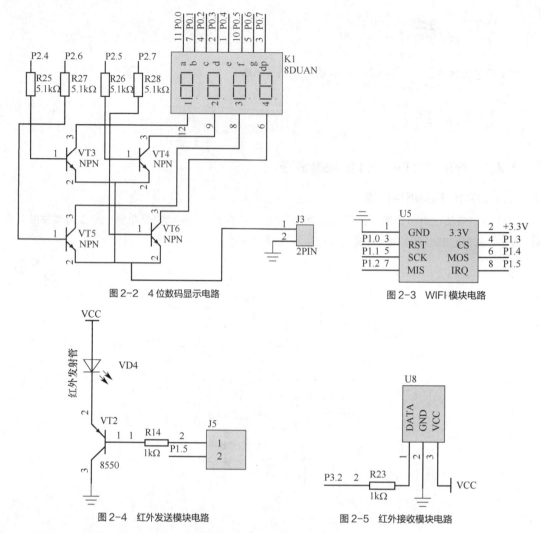

图 2-2　4 位数码显示电路　　　　图 2-3　WIFI 模块电路

图 2-4　红外发送模块电路　　　　图 2-5　红外接收模块电路

2.2.2　亚博科技 BST-M51 学习板功能模块图

亚博科技 BST-M51 学习板功能模块分布如图 2-6 所示。

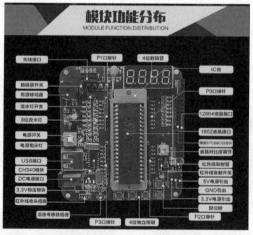

图 2-6　亚博科技 BST-M51 学习板功能模块分布图

Chapter 3

项目3
点亮一个LED

项目目标

- 通过单片机最小系统点亮LED,了解单片机开发程序基础及系统结构和流程。

建议学时

- 8学时。

知识要点

- 单片机系统开发流程。
- C51程序基础
- 进行程序烧录

技能掌握

- 了解发光二极管;建立工程,完成"点亮一个LED";掌握程序的编写、编译,具体需要掌握的内容包括C51的标识符和关键字、常用数据类型、运算符和表达式、程序结构与函数及数组和指针;掌握预处理和程序烧录的方法。

3.1 项目分析

单片机最小系统就是在单片机上接上最少的外围电路元件让单片机工作。在P1.0引脚上输入/输出端口连接1个LED并点亮。通过这一项目将整个开发流程展示出来,包括点亮LED,闪烁LED,控制闪烁时间多项任务。

3.2 技术准备

3.2.1 理论知识

1. LED

LED(Light Emitting Diode)即发光二极管,是半导体二极管的一种,可以把电能转化成光能。LED的图形符号为 ,文字符号记作VD。

2. LED的工作原理

LED与普通二极管一样由1个PN结组成,具有单向导电性。只有当电源正级与LED正极相连,电源负极与LED负极相连时,LED才能导通,发光;反之,LED不能导通,也就不能发光。

3. LED的原理图解析

开发板上LED的原理图如图3-1所示,LED的正极通过一个串联电阻,与电源VCC侧连接,而LED的负极则与单片机的P1端口连接,如果要点亮1支LED就把单片机相对应的I/O端口设为低电平。

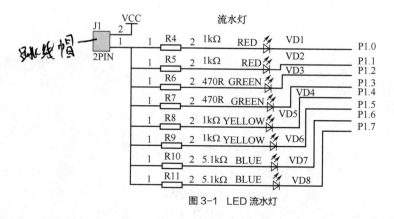

图3-1 LED流水灯

3.2.2 程序基础

1. C51的标识符和关键字

C语言的标识符用来标识源程序中某个对象的名字,这些对象可以是语句、数据类型、函数、变量、数组等。标识符由字符串、数字和下划线等组成,第一个字符必须是字母或下划线。错误的标识符,编译时会有错误提示。由于C51中有些库函数的标识符是以下划线开头的,所以一般不要以下划线开头命名标识符。

例如,f-2、4am、a.m等均为错误标识符。

标准 C 语言定义了 32 个关键字，见表 3-1。

表 3-1　ANSI C 关键字

关　键　字	用　　途	说　　明
auto	存储种类说明	用于声明局部变量，为默认值
break	程序语句	退出最内层循环体
case	程序语句	switch 语句中的选择项
char	数据类型声明	单字节整型数或字符型数据
const	存储种类说明	在程序执行过程中不可修改的值
continue	程序语句	转向下一次循环
default	程序语句	switch 语句中缺省选择项
do	程序语句	构成 do…while 循环结构
double	数据类型声明	双精度浮点数
else	程序语句	构成 if…else 条件结构
enum	数据类型声明	枚举类型数据
extern	存储种类说明	在其他程序模块中声明了的全局变量
float	数据类型声明	单精度浮点数
for	程序语句	构成 for 循环结构
goto	程序语句	构成 goto 循环结构
if	程序语句	构成 if…else 条件结构
int	数据类型声明	整型数
long	数据类型声明	长整型数
register	存储种类说明	使用 CPU 内部寄存器变量
return	程序语句	函数返回
short	数据类型声明	短整型
signed	数据类型声明	有符号整型数
sizeof	运算符	计算表达式或数据类型的字节数
static	存储种类说明	静态变量
struct	数据类型声明	结构体类型数据
switch	程序语句	构成 switch 选择结构
typedef	数据类型声明	重新进行数据类型定义
union	数据类型声明	联合类型数据
unsigned	数据类型声明	无符号数据
void	数据类型声明	无类型数据或函数
volatile	数据类型声明	声明该变量在程序执行中可被隐含地改变
while	程序语句	构成 while 和 do…while 循环结构

C51 在此基础上，针对单片机功能进行了扩展，见表 3-2。

表 3-2 C51 编译器扩充关键字

关 键 字	用 途	说 明
_at_	地址定位	为变量进行绝对地址定位
priority	多任务优先声明	规定 RTX51 或 RTX51 Tiny 的任务优先级
task	任务声明	定义实时多任务函数
alien	函数特性声明	用于声明与 PL/M51 兼容的函数
bdata	存储器类型声明	可位寻址的 MCS-51 内部数据存储器
bit	位变量声明	声明一个位变量或位类型函数
code	存储器类型声明	MCS-51 的程序存储空间
compact	存储器模式	按 compact 模式分配变量的存储空间
data	存储器类型声明	直接寻址 MCS-51 的内部数据寄存器
idata	存储器类型声明	间接寻址 MCS-51 的内部数据寄存器
interrupt	中断函数声明	定义一个中断服务函数
large	存储器模式	按 large 模式分配变量的存储空间
pdata	存储器类型声明	分页寻址的 MCS-5 外部数据空间
sbit	位变量声明	声明一个位变量
sfr	特殊功能寄存器声明	声明一个 8 位特殊功能寄存器
sfr16	特殊功能寄存器声明	声明一个 16 位特殊功能寄存器
small	存储器模式	按 small 模式分配变量的存储空间
using	寄存器组定义	定义 MCS-5 的工作寄存器组
xdata	存储器类型声明	定义 MCS-5 外部数据空间

2. 常量和变量

常量又称为标量,它的值在程序执行过程中不能改变,常量的数据类型有整型、浮点型、字符型和字符串型等。

实际使用中用 #define 定义在程序中经常用到的常量,或者可能需要根据不同的情况进行更改的常量,例如译码地址,而不是在程序中直接使用常量值。这样,一方面有助于提高程序的可读性,另一方面也便于程序的修改和维护。例如:

```
#define PI 3.14                    // 以后的编程中用 PI 代替浮点数常量 3.14,便于阅读
#define SYSCLK 12000000            // 长整型常量用 SYSCLK 代替 12MHz 时钟
#define TRUE  1                    // 用字符 TRUE,在逻辑运算中代替 1
#define STAR '*'                   // 用 STAR 表示字符 "*"
#define uint unsigned int          // 用 uint 代替 unsigned int
```

变量是一种在程序执行过程中,其数值不断变化的量。C51 规定变量必须先定义,后使用。

3. 数据类型

变量都有相应的数据类型,C51 的数据类型见表 3-3。

表 3-3 C51 的数据类型

数 据 类 型	长　　度	值　　域
unsigned char	单字节	0 ~ 255
signed char	单字节	−128 ~ +127
unsigned int	双字节	0 ~ 65 535
signed int	双字节	−32 768 ~ +32 767
unsigned long	4 字节	0 ~ 4 294 967 295
signed long	4 字节	−2 147 483 648 ~ +2 147 483 647
float	4 字节	±1.175 494E−38 ~ ±3.402 823E+38
*	1 ~ 3 字节	对象的地址
bit	位	0 或 1
sfr	单字节	0 ~ 255
sfr16	双字节	0 ~ 65 535
sbit	位	0 或 1

（1）char 字符类型。

char 类型的长度是 1 字节，通常用于定义处理字符数据的变量或常量。分无符号字符类型 unsigned char 和有符号字符类型 signed char，默认值为 signed char 类型。unsigned char 类型用字节中所有的位来表示数值，所能表达的数值范围是 0 ~ 255。signed char 类型用字节中最高位字节表示数据的符号，"0"表示正数，"1"表示负数，负数用补码表示。所能表示的数值范围是 −128 ~ +127。unsigned char 常用于处理 ASCII 字符或用于处理 ≤ 255 的整型数。

（2）int 整型。

int 整型长度为 2 字节，用于存放 1 个双字节数据。分有符号 int 整型数 signed int 和无符号整型数 unsigned int，默认值为 signed int 类型。signed int 表示的数值范围是 −32 768 ~ +32 767，字节中最高位表示数据的符号，"0"表示正数，"1"表示负数。unsigned int 表示的数值范围是 0 ~ 65 535。

（3）long 长整型。

long 长整型长度为 4 字节，用于存放 1 个 4 字节数据。分有符号 long 长整型 signed long 和无符号长整型 unsigned long，默认值为 signed long 类型。signed int 表示的数值范围是 −2 147 483 648 ~ +2 147 483 647，字节中最高位表示数据的符号，"0"表示正数，"1"表示负数。unsigned long 表示的数值范围是 0 ~ 4 294 967 295。

（4）float 浮点型。

float 浮点型在十进制中具有 7 位有效数字，是符合 IEEE-754 标准的单精度浮点型数据，占用 4 字节。

（5）指针型。

指针型是一种特殊的数据类型，其本身就是一个变量，但在其中存放的是另一个数据的地址。在 C51 中，指针的长度一般为 3 字节。根据所指向的变量类型的不同指针变量也有不同的类型，指针变量的类型也就表示了该指针指向的地址中的数据的类型。

（6）bit 位标量。

bit 位标量是 C51 的一种扩充数据类型，利用它可定义一个位标量，但不能定义位指针，也不能定义位数组。它的值是一个二进制位，不是 0 就是 1，位变量的值可以取 0（假）或 1（真）。对位变量进

行定义的语法为
```
bit flag1;
bit send_en=1;
```
（7）sfr 特殊功能寄存器。

单片机内的各种控制寄存器、状态寄存器以及 I/O 端口锁存器、定时器、串行端口数据缓冲器是内部数据存储器的一部分，离散地分布在 80H ~ FFH 的地址空间范围内，这些寄存器统称特殊功能寄存器（Special Function Registers，SFR）。

sfr 类型的长度为 1 字节，其定义方式为
```
sfr 特殊功能寄存器名 = 地址常量；
```
说明"地址常量"就是所定义的特殊功能寄存器的地址，举例如下。
```
sfr TMOD=0x89;          /*定义定时器/计数器方式控制寄存器 TMOD 的地址为 89H*/
sfr P1=0x90;            /*定义 P1 口的地址为 90H*/
```
注意，在关键字 sfr 后面必须是一个名字，名字可以任意选取，但应符合一般的习惯。等号后面必须是常数，不允许有带运算符的表达式，而且该常数必须在特殊功能寄存器的地址范围之内（80H ~ 0FFH）。

（8）sfr16 16 位特殊功能寄存器。

在新一代的 8051 单片机中，特殊功能寄存器在功能上经常组合成 16 位来使用。为了有效地访问这种 16 位的特殊功能寄存器，可采用关键字 sfrl6。sfrl6 类型的长度为 2 字节，其定义语法与 8 位 SFR 相同，但 16 位 SFR 的低端地址必须作为 sfr16 的定义地址，举例如下。
```
sfr16 T2=0CCH;    // 定义 TIMER2，其地址为 T2L=0CCH，T2H=0CDH。
```
（9）sbit 可寻址位。

sbit 同样是 C51 中一种扩充数据类型，利用它能访问芯片内部的 RAM 中的可寻址位或特殊功能寄存器中的可寻址位，举例如下。

PSW 是可位寻址的 SFR，其中各位可用 sbit 定义。
```
sbit    CY=0Xd7;        /*定义进位标志 CY 的地址为 D7H*/
sbit    AC=0xD0^6;      /*定义辅助进位标志 AC 的地址为 D6H*/
sbit    RS0=0XD0^3;     /*定义 RS0 的地址为 D3H*/
```
注意，sfr 和 sbit 只能在函数外使用，一般放在程序的开头。

实际上大部分特殊功能寄存器及其可位寻址的位的定义在 reg51.h、reg52.h 等头文件中已经给出，使用时只需在源文件中包含相应的头文件，即可使用 SFR 及其可位寻址的位；而对于未定义的位，使用前必须先定义，举例如下。
```
#include<reg51.h>
sbit        P10=P1^0;
sbit        P12=P1^2;
main()
{
  P10=1;
  P12=0;
  PSW=0x08;
  ……
}
```

4. C51 运算符和表达式

C51 的运算符主要有算术运算符、关系运算符、逻辑运算符、赋值及复合赋值运算符等。

（1）C51 最基本的算术运算符有以下 5 种。

① "+" 加法运算符。

② "-" 减法运算符。

③ "*"乘法运算符。
④ "/"除法运算符。
⑤ "%"模运算或取余运算符。

对于除法运算符：若2个整数相除，结果为整数（即取整）。

对于取余运算符：要求"%"两侧的操作数均为整型数据，所得结果的符号与左侧操作数的符号相同。

（2）自增、自减运算符。"++"为自增运算符，"--"为自减运算符。如 ++j，j++，--i，i--。

"++"和"--"运算符只能用于变量，不能用于常量和表达式。++j 表示先加1，再取值；j++ 表示先取值，再加1。同理，自减运算也是这个道理。

（3）算术表达式和运算符的优先级与结合性。用算术运算符和括号将运算对象连接起来的式子称为算术表达式。其中的运算对象包括常量、变量、函数、数组、结构等。如 a+b*c/d。

C51 规定算术运算符的优先级和结合性为，先乘、除、模，后加、减，括号最优先。

如果一个运算符两侧的数据类型不同，则必须通过数据类型转换将数据转换成同种类型。转换方式有以下2种。

① 自动类型转换，即在程序编译时，由C编译器自动进行数据类型转换。转换规则如图 3-2 所示。

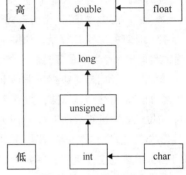

图 3-2　数据类型的转换

一般来说，当运算对象的数据类型不相同时，先将较低的数据类型转换成较高的数据类型，运算结果为较高的数据类型。

② 强制类型转换，使用强制类型转换运算符，其形式为

（类型名）（表达式）

例如：

(double)a;　/*将a强制转换成double类型*/
(int)(x+y);　/*将x+y强制转换成int类型*/

（4）关系运算符和关系表达式。

关系运算实际上就是"比较运算"，将2个表达式进行比较以判断是否和给定的条件相符。

关系运算符包括"<（小于）""<=（小于等于）"">（大于）"">=（大于等于）""==（等于）""!=（不等于）"。

优先级关系是"<""<="">"">="这4个运算符的优先级相同，处于高优先级；"=="和"!="这2个运算符的优先级相同，处于低优先级。关系运算符的优先级低于算术运算符的优先级，而高于赋值运算符的优先级。

用关系运算符将运算对象连接起来的式子称为关系表达式，如 a>b，a+b>=c+d,（a=3）<（b=2）等。关系表达式的结果有2种，即1（真）或0（假）。

（5）逻辑运算符和逻辑表达式。

逻辑运算是对逻辑量进行运算。C51 提供3种逻辑运算符，包括"&&（逻辑与）""||（逻辑或）""!（逻辑非）"。

它们的优先级关系是"!"的优先级最高，而且高于算术运算符；"||"的优先级最低，它低于关系运算符，却高于赋值运算符。

用逻辑运算符将运算对象连接起来的式子称为逻辑表达式。运算对象可以是表达式或逻辑量，而表达式可以是算术表达式、关系表达式或逻辑表达式。逻辑表达式的值也是逻辑量，即"真"或"假"。对于算术表达式，其值若为0，则认为是逻辑假；若不为0，则认为是逻辑"真"。逻辑表达式的执行规

则是，逻辑表达式不一定完全被执行，只有当一定要执行下一个逻辑运算符才能确定表达式的值时，才执行该运算符。

例如：a&&b&&c

若 a 的值为 0 则不需要判断 b 和 c 的值就可确定表达式的值为 0。

又如：a||b||c

若 a 值为 0，则还需判断 b 的值，若 b 的值为 1，则不需要判断 c 的值就可确定表达式的值为 1。

（6）位运算符及其表达式。

位运算的操作对象只能是整型和字符型数据，不能是实型数据。C51 提供以下 6 种位运算。

① "&" 按位"与"，相当于 ANL 指令。

参加运算的 2 个运算量，如果 2 个相应的位都是 1，则结果值中的该位为 1，否则为 0。主要用于清除或者取出一个数中的某些特定位。

② "|" 按位"或"，相当于 ORL 指令。

参加运算的 2 个运算量，如果 2 个相应的位至少有 1 个是 1，则结果值中的该位为 1，否则为 0。按位"或"运算常用来对一个数据的某些特定位置 1。

③ "^" 按位"异或"，相当于 XRL 指令。

参加运算的 2 个运算量，如果 2 个相应的位相同，即均为 1 或均为 0，则结果值中的 i 位为 0，否则为 1。

按位"异或"运算常用来对一个数据的某些特定位进行翻转。

④ "~" 按位取反，相当于 CPL 指令。

"~"是一个单目运算符，用来对一个二进制数按位取反，即将 0 变 1，将 1 变 0。

⑤ "<<" 左移，相当于 RL 指令。

左移运算符用来将一个数的各二进制位全部左移若干位，移到左端的高位被舍弃。最低位补 0。左移 1 位相当于乘以 2，左移 n 位相当于乘以 2^n。

⑥ ">>" 右移，相当于 RR 指令。

右移运算符用来将一个数的各二进制位全部右移若干位，移到右端的低位被舍弃。对无符号数或者有符号数中的正数，左边高位移入 0；对有符号数中的负数，左边高位移入 1。

右移 1 位相当于除以 2，右移 n 位相当于除以 2^n，因此 a>>2 相当于 a/4。

（7）赋值运算符和赋值表达式。

① 赋值运算符。

赋值运算符就是赋值号"="，赋值运算符的优先级低，结合性是右结合性。

② 赋值表达式。

将一个变量与表达式用赋值号连接起来就构成赋值表达式。赋值表达式包括变量、算术运算表达式、关系运算表达式、逻辑运算表达式等，甚至可以是另一个赋值表达式。赋值过程是将"="右边表达式的值赋给"="左边的一个变量，赋值表达式的值就是赋值变量的值。

例如：　　　　　　　　a=b=5，该表达式的值为 5

　　　　　　　　a=（b=4）+（c=6），该表达式的值为 10

③ 赋值的类型转换规则。在赋值运算中，当"="两侧的类型不一致时，系统自动将右边表达式的值转换成左边变量的类型，再赋给该变量。转换规则如下。

a. 实型数据赋给整型变量时，舍弃小数部分。

b. 整型数据赋给实型变量时，数值不变，但以浮点数形式存储在变量中。

c. 长字节整型数据赋给短字节整型变量时，实行截断处理。如将 long 型数据赋给 int 型变量时，将 long 型数据的低 2 字节数据赋给 int 型变量,而将 long 型数据的高 2 字节的数据丢弃。

d. 短字节整型数据赋给长字节整型变量时,进行符号扩展。如将 int 型数据赋给 long 型变量时,将 int 型数据赋给 long 型变量的低 2 字节,而将 long 型变量的高 2 字节的每一位都设为 int 型数据的符号值。

(8)复合赋值运算符。

赋值号前加上其他运算符构成复合运算符。C51 提供以下 10 种复合运算符,即 " + = "" – = "" * = "" / = "" % = "" & = "" | = "" ^ = "" << = "" >> = "。等价于先进行运算符后,再进行赋值操作,举例如下。

     a+=b   等价于 a=(a+b)
     x*=a+b 等价于 x=(x*(a+b))
     a&=b   等价于 a=(a&b)
     a<<=4   等价于 a=(a<<4)

(9)逗号运算符与逗号表达式。

C51 提供了一种特殊运算符,即逗号运算符,用逗号运算符可以把 2 个或多个表达式连接起来,形成逗号表达式。逗号表达式的一般形式为

表达式 1,表达式 2,……,表达式 n

逗号表达式的求解过程是从左到右依次计算出每个表达式的值,整个逗号表达式的值等于最右边的表达式(表达式 n)的值。

(10)条件运算符与条件表达式。

条件运算符" ? : "是唯一的一个三目运算符,条件表达式的一般形式为

逻辑表达式? 表达式 1: 表达式 2

条件表达式的求解过程是首先计算逻辑表达式的值,如果为 1(真),则整个表达式值为表达式 1 的值,否则为表达式 2 的值。

5. C51 程序结构

C51 程序与其他语言程序一样,程序结构也分为顺序结构、选择结构或分支结构、循环结构 3 种。由于顺序结构比较简单,在此不多讲述,下面就选择语句和循环语句进行叙述。

(1)选择语句。

选择语句即条件判断控制语句,它首先判断给定的条件是否满足,然后根据判断的结果决定执行给出的若干种选择之一。C51 中选择语句有 if 语句、switch/case 语句。

① if 语句。

C51 提供如下 3 种形式的 if 语句。

a.

```
if(表达式)
{语句;}
```

示例如下。

```
if(p1!=0)
  {c=20;}
```

b.

```
if(表达式)
{语句 1;}
else
{语句 2;}
```

示例如下。

```
if(p1!=0)
  {c=20;
```

```
Else
    {c=0;}
```
C.
```
if（表达式1）{语句1;}
else if（达式2）{语句2;}
else if（表达式3）{语句3;}
……
else if（表达式n）{语句n;}
else {语句n+1;}
```
示例如下。
```
if（a>=1）{c=10;}
else if（a>=2）{c=20;}
    else if（a>=4）{c=40;}
        else {c=0;}
```
在if语句中又含有1个或多个if语句，这种情况称为if语句的嵌套。If语句嵌套的基本形式如图3-3所示。

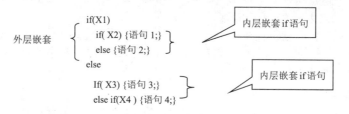

图3-3 if语句嵌套的基本形成

请注意if与else的对应关系，else总是与它前面最近的一个if语句相对应，最好使内层嵌套的if语句也包含else部分（不要省略），这样，程序中if的数目与else的数目一一对应，不至于出错。另外在编程时最好使用相同深度的缩进排写的形式将同一层次上的if-else语句在同一列的位置上对齐，这样不仅不易出错，而且便于程序阅读。

② switch/case语句。

switch/case语句是多分支选择语句，一般形式为

```
switch     (表达式)
    {
    case 常量表达式1: 语句1;break;
    case 常量表达式2: 语句2;break;
    ……
    case 常量表达式n: 语句n;break;
    default: 语句n+1;
    }
```

当switch括号中的表达式的值与某一case后面的常量表达式的值相同时，就执行它后面的语句，然后因遇到break而退出switch语句。若所有的case中的常量表达式的值都没有与表达式的值相匹配时，就执行default后面的语句。

每一个case的常量表达式必须是不相同的，否则将出现混乱局面。

各个case和default出现的次序，不影响程序的执行结果。

如果在case语句中遗忘了break，则程序在执行本行之后，不会按规定退出switch语句，而是执行后续的case语句。

（2）循环语句。

C51循环分当型循环和直到型循环2种，循环实现方法有如下4种。

① if语句和goto语句。

goto 语句只能构成简单循环，与 if 语句一起可以实现当型和直到型循环。

a. 构成当型循环。

```
loop: if（表达式）
         { 语句
           goto loop;
         }
```

b. 构成直到型循环。

```
loop:{ 语句
         if（表达式）goto loop;
      }
```

② while 语句。

while 语句用来实现当型循环。其一般格式为

```
while（表达式）语句
```

表达式可以是任何表达式，语句可以是复合语句。

while 语句的执行过程为，计算表达式的值，当表达式的值为非 0 时，执行内嵌语句（循环），当表达式的值为 0 时，退出 while 循环。

③ do…while 语句。

do…while 语句实现直到型循环。其一般格式为

```
do 语句 while（表达式）;
```

do…while 语句的特点是，先执行语句，后判断表达式。执行过程为，执行内嵌的语句，计算表达式，当表达式的值为非 0 时，循环；当表达式的值为 0 时，执行 do…while 语句下面的语句。

④ for 语句。

for 语句的一般形式为

```
for（表达式 1；表达式 2；表达式 3）语句
```

其执行过程如下。

a. 求解表达式 1。

b. 求解表达式 2，当表达式的值为非 0 时，执行内嵌语句；当表达式的值为 0 时，退出循环。

c. 求解表达式 3，回到（b）。

for 语句最简单的应用形式为

```
for（循环变量初值；循环条件；循环变量改变）语句
```

6. C51 函数

C 程序由一个主函数 main（ ）和若干个其他函数组成。主函数可调用其他函数，其他函数也可以互相调用，同一个函数可以被调用多次。

（1）函数的定义。

函数定义的一般形式为

```
返回值类型 函数名（形式参数列表）
{
    函数体
}
```

例如，int max（int x, int y, int z）

返回值的数据类型为整型；函数名为 max；x，y，z 为 3 个整型入口参数。

① 关于返回值类型有下面几种情况。

a. 可以是基本数据类型（int，char，float，double）及指针类型。

b. 当函数没有返回值时，用标识符 void 说明该函数没有返回值。

c. 若没有指定返回值类型，默认返回值类型为整型类型。

d. 1个函数只有1个返回值，该返回值是通过函数中的return语句获得的。

② 函数名必须是一个合法的标志符。

③ 形式参数列表包括了函数所需全部参数的定义。此时函数的参数称为形式参数，简称形参。形参可以是基本数据类型的数据、指针类型数据、数组等。在没有调用函数时，函数的形参和函数内部的变量未被分配内存单元，即它们是不存在的。

④ 函数体由两部分组成，即函数内部变量定义和函数体其他语句。

注意，各函数的定义是独立的；函数不能定义在另一个函数的内部。

（2）函数返回值。

① 返回语句return用来回送一个数值给定义的函数，从函数中退出。

② 返回值是通过return语句返回的。

③ 返回值的类型如果与函数定义的类型不一致，那么返回值将被自动转换成函数定义的类型。

④ 如果函数无须返回值，可以用void类型说明符指明函数无返回值。

（3）形式参数与实际参数。

与使用变量一样，在调用一个函数之前，必须对该函数进行声明。函数声明的一般格式为

函数类型 函数名（形式参数列表）

函数定义时参数列表中的参数称为形式参数。函数调用时所使用的替换参数，是实际参数，简称实参。定义的形参与函数调用的实参类型应该一致，书写顺序应该相同。函数调用的一般形式为

函数名（实际参数列表）；

在一个函数中需要用到某个函数的功能时，就调用该函数。调用者称为主调函数，被调用者称为被调函数。若被调函数是有参函数，则主调函数必须把被调函数所需的参数传递给被调函数。传递给被调函数的数据称为实际参数。若被调函数是无参函数，则调用该函数时，可以没有参数列表，但括号不能省。被调函数执行完后再返回主调函数继续执行剩余程序。实参与形参在数量、类型和顺序上都必须一致；实参可以是常量、变量和表达式；实参对形参的数据传递是单向的，即只能将实参传递给形参。

（4）调用函数的方式。

① 被调用的函数必须是已经存在的函数。调用函数的方式如下。

函数作为语句。把函数调用作为一个语句，不使用函数返回值，只是完成函数所定义的操作，示例如下。

```
refresh_led( );
```

② 函数作为表达式。函数调用出现在一个表达式中，使用函数的返回值，示例如下。

```
int k;
k=sum(a,b);
```

③ 函数作为一个参数。函数调用作为另一个函数的实参，示例如下。

```
int k;
k=sum(sum(a,b),c);
```

7. C51构造数据类型

前面介绍的整型、字符型、浮点型属于基本数据类型。还有一些数据类型是以上基本数据类型经组合封装而成，称为构造数据类型。C51与标准C语言类似，它的常用构造数据类型是数组和指针。此外还有结构体（Struct）、共用体（Union）和枚举（Enum）等，请大家参阅有关资料，就不在此介绍了。

（1）一维数组。

① 数组是关键数据的有序集合，数组中的每个元素都是统一类型的数据。数组集合用一个名字来标识。数组中元素的顺序用下标表示，下标表示该元素在数组中的位置。下标为n的元素可以表示为数组名[n]。改变[ ]中的下标就可以访问数组中所有的元素。由具有一个下标的数组元素组成的数组称为

一维数组。

② 一维数组的定义。一维数组定义的一般形式为

类型说明符数组名 [元素个数]；

数组名是一个标识符，元素个数是一个常量表达式，不能是含有变量的表达式，示例如下。

```
int a[50];          /** 定义一个数组名为 a 的数组，数组包含 50 个整型的元素 */
```

③ 一维数组的初始化。在定义数组时可以对数组整体初始化，若定义后要对数组赋值，则只能对每个元素分别赋值。示例如下。

```
int a[5]={1,2,3,4,5};   /*给全部元素赋值，a[0]=1，a[1]=2，a[2]=3，a[3]=4，a[4]=5*/
int b[6]={1,2,6};       /*给部分元素赋值，b[0]=1，b[1]=2，b[2]=6，b[3]=b[4]=b[5]=0*/
int d[10];d[0]=4;d[1]=-6;   /*定义完后再赋值*/
```

C 语言不允许对数组的大小作动态定义，示例如下。

```
int i=5;
    int a[i];  /*本语句错误*/
```

（2）二维数组。

由具有 2 个下标的数组元素组成的数组称为二维数组。

① 二维数组的定义。二维数组定义的一般形式为

类型说明符数组名 [行数][列数]；

数组名是一个标识符，行数和列数都是常量表达式，示例如下。

```
float a[3][4];      /*a 数组有 3 行 4 列共 12 个实型元素 */
```

② 二维数组的初始化。

与一维数组的初始化相似，定义时可以整体初始化，也可以在定义后，进行单独赋值，示例如下。

```
int a[3][4]={{1,2,3,4},{5,6,7,8},{9,10,11,12}};          /*全部初始化 */
```

（3）字符数组。

若一个数组的元素是字符型的，则该数组就是字符数组。字符数组的定义与赋值，与一维数组的定义赋值的方法类似，示例如下。

```
char a[12]={ "Chong Qing"};   /*将字符数组初始化为 Chong Qing*/
```

C 语言中没有字符串变量，需用字符数组来处理字符串。当数组中存放的实际字符个数与数组的长度不相同时，为了测定字符串的实际长度和使用系统提供的各种字符串函数，C 语言规定了字符串结束标志"\0"，它是一个 ASCII 码值为 0 的字符。在一个字符数组中，一旦遇到字符"\0"，就表示字符串结束，其后的字符忽略不计。上面 a 数组的后 2 个元素皆为"\0"，字符串常量中也自动包含一个字符串结束符"\0"。

如果提供的初值个数与预定的数组长度相同，在定义的时候可以忽略数组长度，系统会自动根据初值个数确定数组长度，示例如下。

```
static  char  a[]="Welcome you!"
```

数组 a 的长度自动定为 12。

（4）查表。

数组的一个很重要的用途就是查表。在单片机应用中，常常要对数学公式进行计算以及对一些传感器的非线性进行补偿，这时，采用查表的办法比较简单有效。因为单片机的计算能力有限，可以将复杂的数学公式或补偿算法事先计算成表格，存入程序存储器中，而这个表格就是数组。

（5）指针。

前面数据类型已经说过指针变量是一种特殊的变量，特殊在它只能存放地址值。指针在 C 语言中是非常重要的概念。

① 指针运算符。

a."&" 取地址运算符。

当把某变量的地址赋给一个指针变量后，就称该指针变量指向该变量。

例如，

```
int a, *pointer,b;
pointer =&a;  // 指针变量pointer指向变量
```

b。"*" 取指向运算符，即取指针变量所指向的变量值。

例如，b=*pointer；=> b=a；

② 指针类型。

当定义一个指针变量时，若未指定它所指向的对象的存储器类型，则该指针变量被认为是一般指针；反之若指定了它所指向的对象的存储器类型，则该指针被认为是基于存储器的指针。

a. 基于存储器的指针。基于存储器类型的指针由C源代码中指定的存储类型决定，并在编译时确定，这种指针只需1～2字节，并且效率高。基于存储器的指针是在说明一个指针时，指定它所指向的对象的存储类型。一般占2字节，示例如下。char xdata *px；

px为指向一个定义在xdata存储器中的字符变量的指针变量。px本身在默认的存储器区域（由编译模式决定），其长度为2字节。

前面已经讲到，C51有3种存储模式，即SMALL，COMPACT和LARGE。函数的参数与局部变量的存储区域由C51的存储模式确定。在SMALL模式下，函数的参数与局部变量位于单片机的内部RAM；在COMPACT模式下，函数的参数与局部变量位于单片机的外部RAM，示例如下。

```
char xdata *data py；
```

py为指向一个定义在xdata存储器中的字符变量的指针变量。py本身在RAM中，与编译模式无关，其长度也为2字节。

b. 一般指针。一般指针需占3字节，第1个字节为存储器类型的编码（由编译模式的默认值确定），剩余2个字节为地址偏移量。存储器类型决定了对象所用的51单片机的存储空间，偏移量指向实际地址。一个一般指针可以访问任何变量而不管它存储空间的具体位置。这就允许一般函数，如memcpy()等将数据从任意一个地址拷贝到另一个地址空间。

在函数的调用中，函数的指针参数需要用一般指针。一般指针的说明形式为

```
char *pz；
```

这里没有指定指针变量pz所指向的变量的存储类型，pz处于编译模式默认的存储区，长度为3字节。每个字节的含义见表3-4。

表3-4 pz各字节含义

地址	+0	+1	+2
内容	存储类型的编码	高位地址偏移量	低位地址偏移量

其中存储类型由编译器决定，不同的存储区域的编码见表3-5。

表3-5 存储区域编码

存储类型	idata	xdata	pdata	data	code
编码值	1	2	3	4	5

使用常量指针时，如定义外部端口的地址，必须注意正确定义存储类型和偏移量。

例如，将数值0x41写入地址为0x8000外部数据存储器中，可实现语句为

```
#include<absacc.h>
XBYTE[0x8000]=0x41;
```

其中XBYTE是一个指针，是在头文件absacc.h中定义，定义为

```
#define XBYTE ((unsigned char *) 0x20000L)
```

XBYTE被定义为（unsigned char *）0x20000L，是一个一般指针，其存储类型为2，即为xdata型，偏移量是0000。这样，XBYTE成为指向外部数据存储器的零地址单元的指针，而XBYTE[8000]则表示

外部数据存储器的 0x8000 单元。

③ 指针与一维数组。

一维数组在内存中的存放是从下标为 0 的元素开始连续存放的。下标为 0 的元素的地址即为整个数组的地址，示例如下。

```
int  a[4]={0x5678,0x1234,0x5678,0x1234};
```

要用指针变量来处理一维数组，只需将一维数组的地址赋给一个指针变量即可，示例如下。

```
int  *p,a[4];
```

指针的值虽是整型数据，但有其特殊的运算规律。对指针变量，其算术运算的特殊性表现在以下 2 方面。

a. 只能加、减，不能乘、除。

b. 当某指针变量指向一个数组时，对该指针变量加 1，则不管这个数组是何类型，这个指针变量都指向数组的下一个元素（只要没越界）。如做完 p++ 运算后，p 就指向数组 a 的下一个元素，即 a[1]。

数组元素的引用有如下 2 种方法。

a. 下标法。

数组的第 n 个元素可表示为 a[n] 或 p[n]，示例如下。

```
int a[10],*p;p=a;
```

b. 指针法。

数组的第 n 个元素可表示为：*（a+n）或 *（p+n）。

注意，数组名可看成是一个指针常量，其值不能改变。

④ 指针与二维数组。

一个 n 行 m 列的二维数组 a[n][m] 可以看作是一个有 n 个元素的一维数组，其每个元素又是一个具有 m 个元素的一维数组。指向二维数组的指针变量，可以定义指针变量指向二维数组的元素。

⑤ 指针数组和指向指针的指针变量。

a. 指针数组。

与普通变量一样也可以定义指针数组，即每个元素都是同类型的指针变量。常令指针数组指向若干个字符串，这使得字符串的处理更为方便。

其定义的一般形式为

```
类型说明符  *数组名[元素个数];
```

b. 指向指针变量的指针。

指针变量在内存中占用一定的单元，即也有地址。如果一个指针变量用来存放另一个指针变量的地址，这个变量就称为指向指针变量的指针变量。

其定义的一般形式是为

```
类型说明符  **变量名
```

其中，类型说明符是指被指向的指针变量的类型，示例如下。

```
char **P,*p1;
     p=&p1;
```

8. C51 的预处理

预处理功能包括宏定义、文件包含和条件编译 3 个主要部分。预处理命令不同于 C 语言语句，具有以下 3 个特点。

① 预处理命令以"#"开头，后面不加分号。

② 预处理命令在编译前执行。

③ 多数预处理命令习惯放在文件的开头。

（1）宏定义。

宏符号名一般采用大写形式。不带参数的宏定义的格式为

```
#define 宏符号名常量表达式
```

例如：#define PI 3.14   /* 用宏符号名 PI 代替浮点数 3.14。*/
结束宏符号名的定义语句为

```
#undef 宏符号名
```

（2）包含文件。

包含文件的含义是在一个程序文件中包含其他文件的内容。用文件包含命令可以实现文件包含功能，命令格式为

```
#include< 文件名 > 或 #include "文件名"
例如，在文件 file1.c 中
#include "file2.c"
main(){
  ...
}
```

在编译预处理时，对 #include 命令进行文件包含处理，实际上就是将文件 file2.c 中的全部内容复制插入到 #include "file2.c" 的命令处。

（3）条件编译命令。

提供一种在编译过程中根据所求条件的值有选择地包含不同代码的手段，实现对程序源代码的各部分有选择地进行编译，称为条件编译。

#if 语句中包含一个常量表达式，若该表达式的求值结果不等于 0 时，则执行其后的各行，直到遇到 #endif，#elif 或 #else 语句为止（预处理 elif 相当于 else if）。在 #if 语句中可以使用一个特殊的表达式 defined（标识符），当标识符已经定义时，其值为 1；否则，其值为 0。

例如，为了保证 hdr.h 文件的内容只被包含一次，可把该文件的内容包围起来的条件语句为

```
#ifndef (HDR)
#define HDR
#include (hdr.h)
#endif
```

3.3 项目实施

显然，从 LED 原理图上看，只要 P1.0 引脚输出为低电平就可以点亮 LED。

软件设计不用汇编，一律采用 C51 语言，便于理解和扩展。点亮第 1 只 LED 的实验程序如下。

```
/*********************************
*实验效果：程序烧录进去后，第1只LED被点亮
*********************************/
#include<reg52.h>
//#define led P1
sbit led1 = P1^0;
void main()
{
  while (1)
  {
    led1 = 0;// 点亮第1个LED
    //led = 0xfe;//1111 1110    最低位 P1^0 = 0;
    //led4 = 0;
  }
}
```

程序的第 1 行是"文件包含"，是将另一个文件"reg52.h"的内容全部包含进来。文件"reg52.h"

包含了 51 单片机全部的特殊功能寄存器的字节地址及可寻址位的位地址定义。用 #include "at89x52.h" 也可以，其具体的区别在于，后者定义了更多的地址空间。可找到相应的头文件，进行比较。

本程序包含 reg52.h 的目的就是为了使用 P1 这个符号，即通知程序中所写的 P1 是指 AT89S52 的 P1 端口，而不是其他变量。

打开 reg52.h 文件可以看到 "sfr P1=0x90；"，即定义符号 P1 与地址 0x90 对应，而 P1 端口的地址就是 0x90。虽然这里的 "文件包含" 只有 1 行，但 C 编译器在处理的时候却要处理几十行或几百行。

程序的第 3 行用符号 "led1" 表示 P1.0 引脚。在 C51 中，如果直接写 "P1.0" 编译器并不能识别，而且 "P1.0" 也不是一个合法的 C51 语言程序变量名，必须重命名，此处重命名为 "led1"，但 "led1" 与 "P1.0" 必须建立联系，这里使用了 C51 的关键字 "sbit" 来进行定义。

"while（1）" 语句使单片机工作在死循环状态，目的是一直输出低电平。

把接在单片机 P1_0 上的 LED 点亮了，此电路中是低电平点亮 LED，因为已经把 LED 的阳极接至 VCC。

1. 使用 Keil C51 编写程序

参考项目一的使用 Keil 的方法，将该程序编写出来，如图 3-4 所示。

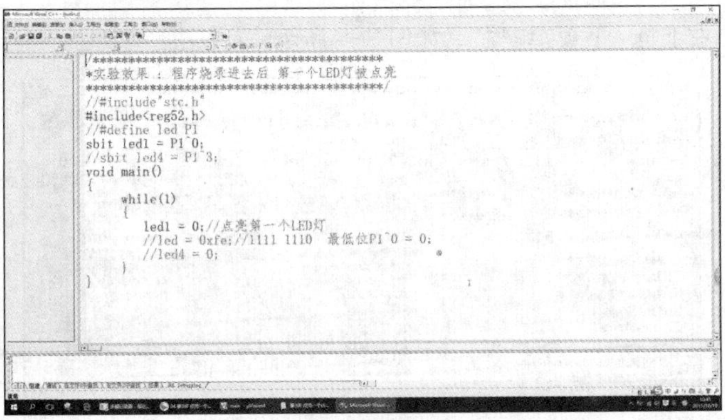

图 3-4　编写程序

2. 烧录程序

打开烧录软件，并且选择好单片机型号和串口，如图 3-5 所示。

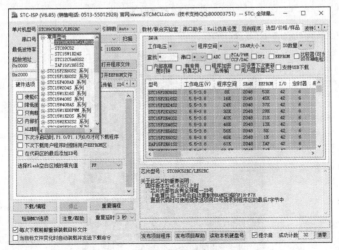

图 3-5　烧录程序

单击"打开程序文件",打开之前编写好的程序(.hex 文件),如图 3-6 所示。

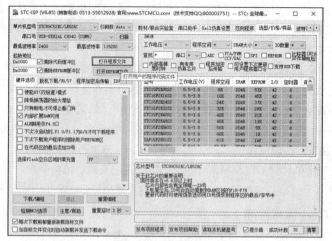

图 3-6　打开程序文件

程序烧入,单击右下角"下载/编程",如图 3-7 所示。

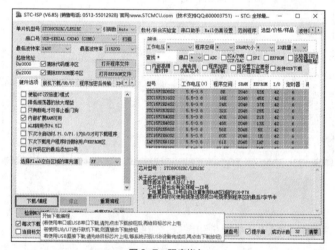

图 3-7　程序烧入

点亮 LED 实验结果实物图如图 3-8 所示。

图 3-8　点亮 LED 实验结果实物图

Chapter 4

项目 4
流水灯实验

项目目标

- 通过流水灯控制来了解C51程序。

建议学时

- 4学时。

知识要点

- C语言知识点,包括宏定义,包括函数和函数调用、循环结构、while语句、关系运算和逻辑值。
- 编流程图绘制
- 闪烁灯设计。
- 函数调用。

技能掌握

- 常用C51程序设计。

4.1 项目分析

前面学习了如何点亮 LED 灯,现在扩展到构成最简单的流水灯。本项目完成一个典型流水灯的设计,便于学生掌握 C51 语言的数据类型、程序结构、函数、数组等基本概念。

4.2 技术准备

4.2.1 流水灯硬件实物

流水灯是由多个 LED 灯组成的,如图 4-1 所示。

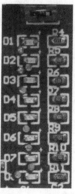

图 4-1 流水灯

4.2.2 流水灯原理图

编流水灯程序之前,必须先读懂单片机上的流水灯原理图,如图 4-2 所示。

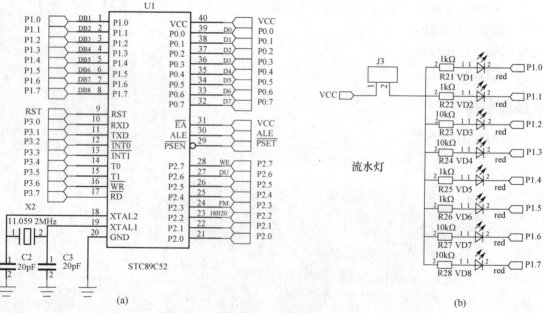

图 4-2 流水灯原理图

4.2.3 流水灯实验理论知识

1. 延时函数

每条汇编指令的进行都要占用一定的时间（或者机器周期），如果让机器什么都不做，即空指令的话，机器就会延时，然后在计算好每次延时到底有多长，外面套一个循环（或者多重循环），根据想要的延时时间即可计算出来循环的次数，延时函数的参数就是用来控制循环次数的。

实现延时通常有 2 种方法：一种是硬件延时，要用到定时器/计数器，这种方法可以提高 CPU 的工作效率，也能做到精确延时；另一种是软件延时，这种方法主要采用循环体进行延时。

2. 软件延时

使用定时器/计数器可实现精确延时。但在很多情况下，定时器/计数器被用于其他用途，这时候就只能用软件方法实现较精确的延时。软件延时与时间计算的方法如下。

① 短暂延时。
② 在 C51 中嵌套汇编程序段实现延时。
③ 使用示波器确定延时时间。
④ 使用反汇编工具计算延时时间。

3. For 循环语句

```
For（表达式 1；表达式 2；表达式 3）
{ 语句（内部可为空）}
```

执行过程如下。

（1）求解 1 次表达式 1。

（2）求解表达式 2，若其值为真（非 0），则执行 for 中语句，然后执行（3）。否则结束 for 语句，直接跳出，不再执行（3）。

（3）求解表达式 3。

（4）跳到（2）重复执行。

4. 一个简单的延时函数

只需要了解，并学会调用它，不需全记住。

```
/*------------------------------------------------
延时函数，含有输入参数 unsigned int t，无返回值
  unsigned int 是定义无符号整形变量的，其值范围是
  0～65 535
------------------------------------------------*/
void Delay(unsigned int t)
  {
  while(--t);
  }
```

5. 宏定义

宏定义"#define"中"#"号是预处理指令，用在开头。宏定义应用举例如下。

```
#define ON（新的名称）1（原来的名称）
```

宏定义的作用是在程序书写时可用新的名称代替原来的名称书写，以达到简化或直观的效果。编译器在预处理的时候，自动将新的名称还原为原来的名称。

宏定义的书写位置在源程序开头，函数的外面。

6. 3 个流水灯程序框图

3 个流水灯程序有各自的流程，流水灯程序框图如图 4-3 所示。

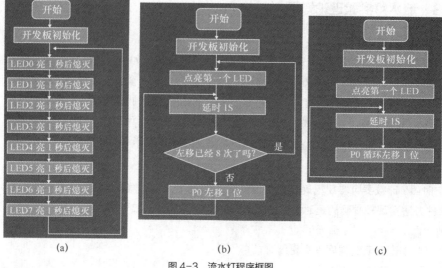

图 4-3 流水灯程序框图

7. 函数的引入

函数是 C 语言的主要特点,也是学习的一个重点。

主函数只有一个。

在实际编程中,会遇到以下 2 种情况。

(1)使用已有程序,比如网络共享资源,如 DS18B20 功能函数。

(2)会反复使用某一段程序,这就需要用到函数和利用函数调用其他功能函数。

注意,函数一般是指功能函数。

8. While 循环语句

```
While(关系表达式或逻辑值);
While(关系表达式或逻辑值)一条语句;
While(关系表达式或逻辑值)(注意:这里没有分号)
    {
    语句 1;
    语句 2;
    语句 3;
    ……
    语句 n;
    }
```

特点:先判断条件表达式,后执行语句。

原则:若条件表达式的值为真,那么执行语句。否则跳出 while 语句。

9. 计算机中的 3 种运算

计算机中有以下 3 种运算。

(1)算术运算:参与运算的对象是数,结果也是数,运算符是加(+)减(-)乘(*)除(/)。

(2)逻辑运算:参与运算的对象是逻辑量,结果也是逻辑量,运算符是与(&&)、或(‖)、非(!)。逻辑值有 2 个,即 1 或非 0(真)和 0(假)。

(3)关系运算:参与运算的对象是数,结果是逻辑量,运算符是大于(>)、小于(<)、大于或等于(>=)、小于或等于(<=)等于(==)(注意,与赋值符"="完全不同)、不等于(!=)。例如关系运算 1<3=1;1==2=0,1+1>2=0。

10. 左移和右移

位运算就是按位对变量进行运算。左移和右移属于位运算。格式为

```
P1<<1;/*左移一位*/
a>>2;/*右移二位*/
```

一般多用于对 8 位无符号数进行移位操作，移出的数丢弃，空位自动补 0。

注意，位运算不改变参与运算的变量的值，如果希望改变参与运算的变量的值，应使用相应的赋值操作。

用左移编写第二个流水灯程序。

4.2.4 定时器的结构

定时器功能由 T0 和 T1，以及他们的工作方式寄存器 TMOD 和控制寄存器 TCON 等组成。内部通过总线与 CPU 相连。定时器 T0 和 T1 各由 2 个 8 位特殊功能寄存器 TH0，TL0 与 TH1，TL1 构成。

工作方式寄存器 TMOD 用于设置定时器的工作模式和工作方式。

控制寄存器 TCON 用于启动和停止定时器的计数，并控制定时器的状态。

定时器的工作方式、启动、停止、溢出标志、计数器等都是可编程的，通过设置寄存器 TMOD，TCON，TH0，TL0，TH1 和 TL1 实现。TH0 和 TL0 存放定时器 T0 的初值或计数结果。TH0 存放高 8 位，TL0 存放低 8 位；TH1 和 TL1 存放定时器 T1 的初值或计数结果。TH1 存放高 8 位，TL1 存放低 8 位。

4.2.5 定时器的 TMOD 和 TCON 寄存器

1. 工作方式寄存器 TMOD

TMOD（89H）格式如图 4-4 所示。

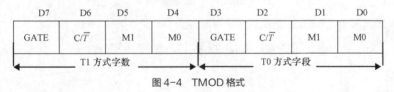

图 4-4　TMOD 格式

（1）GATE 为门控位。

GATE=0 时，定时器的启动不受到外部中断请求信号的影响。一般情况下 GATE=0。

GATE=1 时，T0 的启动受 $\overline{INT0}$ 端口（P3.2 引脚）信号控制，T1 的启动受 $\overline{INT1}$ 端口（P3.3 引脚）信号控制，只有当外部中断信号 $\overline{INT0}$ 和 $\overline{INT1}$ 为高电平的时，才能启动定时器。

（2）M1，M0 为工作方式选择位。

（3）C/$\overline{T}$ 为计数器模式和定时器模式选择位。

C/$\overline{T}$=0 时，定时器模式。

C/$\overline{T}$=1 时，计数器模式。

2. 特殊功能寄存器 TCON

TCON 格式见表 4-1。

表 4-1　TCON 格式

D7	D6	D5	D4	D3	D2	D1	D0
TF1	TR1	TF0	TR0	…	…	…	…

低 4 位与外部中断有关。高 4 位的功能如下。

（1）TF1，TF0 为计数溢出标志位。

定时器 T0 或 T1 计数溢出时，由硬件自动将此位置 1。

TFx 可以由程序查询，也是定时中断的请求源。

（2）TR1，TR0 为计数运行控制位。

TRx=1 时，启动定时器/计数器工作。

TRx=0 时，停止定时器/计数器工作。

4.2.6 定时器工作方式

MCS-51 的定时器 T0 有 4 种工作方式，即方式 0、方式 1、方式 2、方式 3。

MCS-51 的定时器 T1 有 3 种工作方式，即方式 0、方式 1、方式 2。

1. 方式 0

在这种方式下，16 位寄存器 TH1 和 TL1 只用 13 位，由 TH1 的 8 位和 TL1 的低 5 位组成。TL1 的高 3 位不定。

方式 0 定时时间为

$$(2^{13} - 初值) \times 振荡周期 \times 12$$

例如，若晶振频率为 12MHz，则最长的定时时间为 $(2^{13}-0) \times (1/12) \times 12\mu s = 8.191ms$

2. 方式 1

在这种方式下，16 位寄存器 TH1 和 TL1 为 16 位的计数器，除位数外，其他与方式 0 相同。

方式 1 定时时间为

$$(2^{16} - 初值) \times 振荡周期 \times 12$$

例如，若晶振频率为 12MHz，则最长的定时时间为

$$(2^{16}-0) \times (1/12) \times 12\mu s = 65.536ms$$

3. 方式 2

THx 作为常数缓冲器，当 TLx 计数溢出时，在置 1 溢出标志 TFx 的同时，还自动的将 THx 中的初值送至 TLx，使 TLx 从初值开始重新计数。

方式 2 定时时间为

$$(2^{8} - 初值) \times 振荡周期 \times 12$$

例如，若晶振频率为 12MHz，则最长的定时时间为

$$(2^{8}-0) \times (1/12) \times 12\mu s = 0.256ms$$

4. 方式 3

T0 在方式 3 时被拆成 2 个独立的 8 位计数器，即 TH0 和 TL0。

当 T0 处于方式 3 时，T1 仍可设置为方式 0、方式 1 和方式 2。由于 TR1，TF1 和 T1 的中断源都已被定时器 T0 占用，所以定时器 T1 仅有控制位 C/T 来决定其工作在定时方式或计数方式。当计数器计满溢出时，不能置位 TF1，而只能将输出送往串口。所以，此时定时器 T1 一般用作串口的波特率发生器，或不需要中断的场合。

4.2.7 定时器编程步骤

MCS-51 单片机的定时器/计数器是可编程的，具体步骤如下。

（1）对 TMOD 赋值，以确定定时器的工作模式。

计算初值的方法如下。

设计数器的最大值为 M，则置入的初值 X 为

$$\text{计数方式 } X = M - \text{计数值}$$

定时方式由 $(M-X)T=$ 定时值，得 $X=M-$ 定时值 $/T$

T 为计数周期，是单片机的机器周期。

则方式 0 时，$M=2^{13}=8\,192$。

方式 1 时，$M=2^{16}=65\,536$。方式 2 和 3 时，$M=2^{8}=256$。

例如：机器时钟频率为 12MHz，机器周期为 1μs 时。

若工作在方式 0，则最大定时值为 $2^{13} \times 1\mu s = 8.192ms$。

若工作在方式 1，则最大定时值为 $2^{16} \times 1\mu s = 65.536ms$。

（2）置定时/计数器初值，直接将初值写入寄存器的 TH0，TL0 或 TH1，TL1。

（3）对 TCON 寄存器中的 TR0 或 TR1 置位，启动定时/计数器，置位以后，计数器即按规定的工作模式和初值进行计数或开始定时。

（4）查询溢出标志 TFx 的状态，决定是否停止定时/计数器。

4.3 项目实施

注意，J1 接上才能开始做流水灯模块实验。

1. 流水灯模块位转出实验程序

```
/********************************
*实验名：流水灯
*实验效果：流水灯（位输出）
********************************/
#include<reg52.h>
#define uint unsigned int   //16  0-65535
#define on 0
#define off 1

sbit led1 = P1^0;
sbit led2 = P1^1;
sbit led3 = P1^2;
sbit led4 = P1^3;
sbit led5 = P1^4;
sbit led6 = P1^5;
sbit led7 = P1^6;
sbit led8 = P1^7;

void delay(uint xms)// 延时 xms
{
  uint i,j;
  for(i=xms;i>0;i--)      //i=xms 即延时 xms
    for(j=112;j>0;j--);
}

void main()
{
  while(1)
  {
    led1 = on;// P1^0 = 0;
```

```
        delay(200);//延时200ms
        led1 = off;

        led2 = on;
        delay(200);//延时200ms
        led2 = off;

        led3 = on;
        delay(200);//延时200ms
        led3 = off;

        led4 = on;
        delay(200);//延时200ms
        led4 = off;

        led5 = on;
        delay(200);//延时200ms
        led5 = off;

        led6 = on;
        delay(200);//延时200ms
        led6 = off;

        led7 = on;
        delay(200);//延时200ms
        led7 = off;

        led8 = on;
        delay(200);//延时200ms
        led8 = off;
    }
}
```

2. 流水灯模块循环移位实验程序

```
/*********************************
* 实验名：流水灯
* 实验效果：流水灯（循环移位）
**********************************/
#include<reg52.h>
#include<intrins.h>
#include "delay.h"
void main()
{
    P1 = 0xfe;//1111 1110
    while(1)
    {
        delay(500);
        P1 = _crol_(P1,1);/*将P1循环左移1位*/
    }
}
```

3. 移位操作程序

```
/*********************************
* 实验名：流水灯
* 实验效果：流水灯（移位操作）
**********************************/
```

```c
#include<reg52.h>

void delay(unsigned int xms);

void main()
{
    int i=0;
    while(1)
    {
      P1 = 0xfe;//1111 1110
      for(i=0;i<8;i++)
      {
         delay(1000);
         P1 <<= 1;//P1 = P1<<1;
         P1 = P1 | 0x01;//
      }
    }

}

void delay(unsigned int xms)// 延时xms
{
    unsigned int i,j;
    for(i=xms;i>0;i--)              //i=xms 即延时xms
        for(j=112;j>0;j--);
}
```

流水灯实验结果实物图如图 4-5 和图 4-6 所示。

图 4-5　流水灯实验结果实物图 1

图 4-6　流水灯实验结果实物图 2

Chapter 5

项目5
蜂鸣器实验

项目目标

- 通过控制蜂鸣器发声来了解C51的I/O端口。

建议学时

- 4学时。

知识要点

- 放大电路的使用
- 蜂鸣器。

技能掌握

- 学会I/O端口程序编写及放大电路设计。

5.1 项目分析

本项目是单片机最小系统的简单应用。设计一个单片机的最小系统,利用 P1.0 引脚输出电位的变化,控制蜂鸣器发声。P1.0 引脚的电位变化可以通过指令来控制。

5.2 技术准备

5.2.1 蜂鸣器硬件实物

1. 有源蜂鸣器和无源蜂鸣器

蜂鸣器有 2 种,即有源蜂鸣器和无源蜂鸣器。有绿色电路板的一种是无源蜂鸣器,没有电路板而用黑胶封闭的一种是有源蜂鸣器。蜂鸣器实物图如图 5-1 和图 5-2 所示。

图 5-1 无源蜂鸣器

图 5-2 有源蜂鸣器

这里的"源"不是指电源,而是指振荡源。也就是说有源蜂鸣器内部自带振荡源,所以只要一通电就会发出声音。而无源蜂鸣器内部不带振荡源,所以,如果使用直流信号无法令其发出声音,必须使用 2 ~ 5kHz 的方波驱动。

因为有源蜂鸣器内部自带振荡电路,所以价格往往较无源蜂鸣器高,其优点是程序控制方便。相比这下,无源蜂鸣器具有如下优点。

(1)价格便宜。

(2)声音频率可控,可以做出"多来米发索拉西"的效果。

(3)在一些特例中,可以和 LED 复用一个控制口。

2. 蜂鸣器驱动电路

由于蜂鸣器的工作电流通常比较大,单片机的 I/O 端口无法直接驱动,需要通过放大电路即使用三极管放大电流来驱动。

3. 蜂鸣器软件设计方法

I/O 端口电平取反方法,即 I/O 端口的电平进行 1 次翻转,直到蜂鸣器不需要发出声音的时候,将 I/O 端口的电平设置为高电平即可。

5.2.2 蜂鸣器实验相关电路

实验板蜂鸣器电路如图 5-3 所示。

蜂鸣器驱动电路如图 5-4 所示。

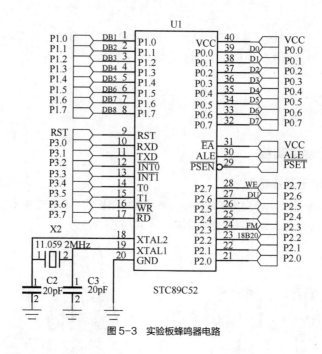

图 5-3　实验板蜂鸣器电路

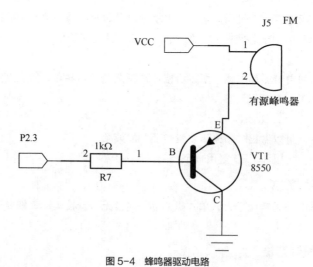

图 5-4　蜂鸣器驱动电路

5.2.3　串行接口的结构

单片机串行接口（简称串口）结构如图 5-5 所示。有 2 个物理上独立的接收、发送缓冲器 SBUF（属于特殊功能寄存器），可同时发送、接收数据。控制寄存器共有 2 个，即特殊功能寄存器 SCON 和 PCON。发送和接收端口（引脚）分别是 TXD（P3.0）和 RXD（P3.1）。

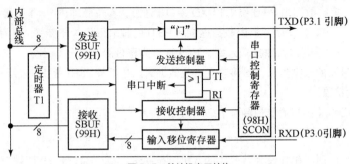

图 5-5 单片机串口结构

1. 串口控制寄存器 SCON

字节地址 98H，可位寻址，位地址为 98H ～ 9FH，格式见表 5-1。

表 5-1 串行口控制寄存器 SCON

位	D7	D6	D5	D4	D3	D2	D1	D0
功能	SM0	SM1	SM2	REN	TB8	RB8	T1	RI
地址	9FH	9EH	9DH	9CH	9BH	9AH	99H	98H

SCON 中各位的功能请参考相关原理书籍，其中 SM0，SM12 位编码所对应的 4 种工作方式见表 5-2。有关 4 种通信方式描述参考 5.2.5。

表 5-2 SM0，SM1 组合设置串行端口的工作方式

SM0	SM1	工作方式	功能简介	比 特 率
0	0	0	移位寄存器	OSC/12
0	1	1	8 位 UART	可变
1	0	2	9 位 UART	OSC/32 或 OSC/64
1	1	3	9 位 UART	可变

注：UART 是一个将并行输入转为串行输出的芯片，集成在单片机内。
OSC 为晶振频率。

2. 特殊功能寄存器 PCON

字节地址为 87H，不能位寻址。格式见表 5-3。

表 5-3 特殊功能寄存器 PCON

位	D7	D6	D5	D4	D3	D2	D1	D0
功能	SMOD	—	—	—	GF1	FG0	PD	IDL

仅最高位 SMOD 与串口有关，SMOD 为波特率选择位。当 SMOD = 1 时，要比 SMOD = 0 时的波特率加倍，所以也称 SMOD 位为波特率倍增位。

5.2.4 串口的 4 种工作方式和波特率

1. 方式 0

方式 0 为同步移位寄存器输入/输出方式。该方式并不用于 2 个 AT89S51 单片机之间的异步串行通信，而是用于串口外接移位寄存器，扩展并行 I/O 端口。

方式0发送时，串行数据由P3.0引脚（RXD端口）送出，移位脉冲由P3.1引脚（TXD端口）送出。在移位脉冲的作用下，串口发送缓冲器的数据逐位从P3.0引脚串行移入外接移位寄存器中。

方式0接收时，RXD端口为数据输入端，TXD端口为移位脉冲信号输出端，接收器以$f_{osc}/12$的固定波特率采样RXD端口的数据信息，当接收完8位数据时，中断标志RI置1，表示一帧数据接收完毕，可进行下一帧数据的接收。

方式0时，SM2位（多机通信控制位）必须为0。

2. 方式1

当SM0，SM1=01时，串口设为方式1的双机串行通信。TXD端口和RXD端口分别用于发送和接收数据，如图5-6所示。

方式1发送时，数据位由TXD端口输出，发送一帧信息为10位：1位起始位0，8位数据位（先低位）和1位停止位1。当CPU执行1条数据写SBUF的指令，就启动发送。发送开始时，内部发送控制信号转换为有效，将起始位向TXD端口（P3.0引脚）输出，此后每经过1个TX时钟周期，便产生1个移位脉冲，并由TXD端口输出1个数据位。8位数据位全部发送完毕后，中断标志位TI置1。

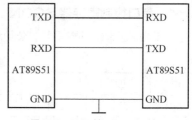

图5-6 TXD端口和RXD端口

方式1接收时（REN=1），数据从RXD端口（P3.1引脚）输入。当检测到起始位的负跳变时，则开始接收。当1帧数据接收完毕后，同时满足以下2个条件，接收才有效。

（1）RI=0，即上1帧数据接收完成时，RI=1发出的中断请求已被响应，SBUF中的数据已被取走，说明"接收SBUF"已空。

（2）SM2=0或收到的停止位=1（方式1时，停止位已进入RB8），则将接收到的数据装入SBUF和RB8（装入的是停止位），且中断标志RI置1。

若不同时满足2个条件，接收的数据不能装入SBUF，该帧数据将被丢弃。

3. 方式2和方式3

方式2和方式3时，串口为9位异步通信接口。每帧数据为11位，1位起始位0，8位数据位（先低位），1位可程控为1或0的第9位数据和1位停止位。除了波特率外，方式3和方式2相同。

（1）方式2发送。

发送前，先根据通信协议由软件设置TB8（如奇偶校验位或多机通信的地址/数据标志位），然后将要发送的数据写入SBUF，即启动发送。TB8自动装入第9位数据位，逐一发送。发送完毕，使TI位置1。

（2）方式2接收。

SM0，SM1=10，且REN=1时，以方式2接收数据。数据由RXD端口输入，接收11位信息。当位检测逻辑采样到RXD端口的负跳变，判断起始位有效，便开始接收1帧信息。在接收完第9位数据后，需满足以下2个条件。

① RI=0，意味着接收缓冲器为空。

② SM2=0或接收到的第9位数据位RB8=1。

才能将接收到的数据送入SBUF（接收缓冲器），第9位数据送入RB8，且RI置1。若不满足这2个条件，接收的信息将被丢弃。

4. 波特率

波特率是串口每秒发送（或接收）的位数。设发送1位所需要的时间为T，则波特率为$1/T$。

定时器的不同工作方式，得到的波特率的范围不一样，这是由T1在不同工作方式下计数位数的不同所决定的。

串行通信，收、发双方发送或接收的波特率必须一致。对于 4 种工作方式，方式 0 和方式 2 的波特率是固定的；方式 1 和方式 3 的波特率是可变的，由 T1 溢出率确定。

在实际设定波特率时，T1 常设置为定时方式 2（自动装初值），即 TL1 为 8 位计数器，TH1 存放备用初值。这种方式操作方便，也可避免因软件重装初值带来的定时误差。

一般波特率可以通过公式计算。本书只介绍常用的波特率，见表 5-4。这里时钟振荡频率选为 11.059 2MHz，就可使初值为整数，从而产生精确的波特率。

表 5-4 常用的波特率

波特率 / kbit/s	f_{osc} / MHz	SMOD 位	方式	初值 X
62.5	12	1	2	FFH
19.2	11.059 2	1	2	FDH
9.6	11.059 2	0	2	FDH
4.8	11.059 2	0	2	FAH
2.4	11.059 2	0	2	F4H
1.2	11.059 2	0	2	E8H

5.2.5　C51 串口编程

单片机的中断系统中第 4 个中断就是串口中断，要进行串口通信首先就要打开 CPU 总中断 EA，还要打开串口通信中断 ES，这是串口通信的前提。

串口通信还要设置 SCON 寄存器来指定采用哪一种方式进行通信。而在通信的过程中，还要设定通信的波特率，否则，单片机无法进行采样。

典型串口例程如下。

（1）Main 函数开始对寄存器进行初始化操作。

例如：

```
TMOD = 0x20;
SCON = 0x50;
TH1 = 0xFA;
PCON = 0x80;
TR1 = 1;
```

（2）发送数据代码（写在程序相应位置上）。

```
SBUF=变量；            // 在相应的位置写此代码，将需要发送的数据送到 SBUF 寄存器中
while（TI==0）;        // 等到数据发送完毕，再执行下一句代码
TI=0;                  //TI 为传送结束标志，必须软件置零
```

（3）接收数据代码（写在程序相应位置上）：

```
while (RI==0);         // 只要接收中断标志位 RI 没有被置 1，直至接收完毕（RI=1）
  RI=0;                // 为了接收下一帧数据，需将 RI 清 0
  变量 =SBUF;          // 将接收缓冲器中的数据存于变量
```

5.3　项目实施

5.3.1　简单蜂鸣器发声实验

单片机的串口通信工作方式有 4 种，需要设置更多的寄存器。前面学习过的定时器与中断是单片机

通信的基础。

蜂鸣器发声实验程序如下。

```c
/*****************************************
*实验名：蜂鸣器实验
*实验效果：程序烧录进去后蜂鸣器发出滴滴的声音
*****************************************/
#include<reg52.h>

#define on 0
#define off 1
sbit fm = P2^3;

void delay(unsigned int xms)
{
   unsigned int i,j;
   for(i=xms;i>0;i--)         //i=xms 即延时 xms
       for(j=112;j>0;j--);
}

void main()
{
    while(1)
    {
        fm = on;//P2^3 = 0; 蜂鸣器响
        delay(100);
        fm = off;
        delay(100);
    }
}
```

5.3.2 给前面任务的流水灯加入报警效果

蜂鸣器配合流水灯发出声音的实验程序如下。

```c
/*****************************************
*实验名：流水灯报警
*实验效果：程序烧录进去后蜂鸣器配合流水灯发出声音
*****************************************/
#include<reg52.h>
#include<intrins.h>

#define on 0
#define off 1
sbit fm = P2^3;

void delay(unsigned int xms)
{
   unsigned int i,j;
   for(i=xms;i>0;i--)         //i=xms 即延时 xms
       for(j=112;j>0;j--);
}

void main()
{
   P1 = 0xfe;//1111 1110
   while(1)
```

```
    {
        delay(100);
        P1 =_crol_(P1,1);/*将P1循环左移1位*/
        fm = on;
        delay(100);
        fm = off;
    }
}
```

蜂鸣器实验结果实物图如图 5-7 所示。

图 5-7 蜂鸣器实验结果实物图

Chapter 6

项目6
LED显示器静态显示

项目目标

- 通过学习在4位LED显示器（数码管）静态显示数字和字母，掌握单片机的输出。

建议学时

- 4学时。

知识要点

- LED显示器的基本类型。
- LED显示器静态扫描。

技能掌握

- 学会LED显示器程序编写及Proteus常用外设和总线图的绘制。

6.1 项目分析

单片机也需要人—机交互,掌握常用的输入和输出设备非常必要。本章选取了最常用的输出模块 LED 显示器来演示其典型程序的编制方法。

6.2 技术准备

6.2.1 LED 显示器静态显示简介

单片机系统中常用的显示器有发光二极管(Light Emitting Diode,LED)显示器、液晶(Liquid Crystal Display,LCD)显示器、CRT 显示器等。LED,LCD 显示器有 2 种显示结构,即段显示(7 段、米字型等)和点阵显示(5×8,8×8 点阵等)。

LED 显示器实物如图 6-1 所示。

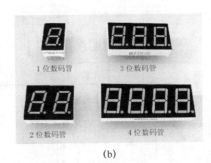

图 6-1 LED 显示器实物

6.2.2 LED 显示器可显示内容和特点

LED 显示器是一种半导体发光器件,其基本单元是发光二极管。LED 显示器可显示的内容包括数字、小数点和部分英文字符、符号。它具有如下 2 个特点。

(1)自发光、亮度高,特别适合环境亮度低的场合使用。

(2)牢固,不怕冲击。

6.2.3 LED 显示器的结构与原理

1. LED 显示器的结构

LED 显示器系发光器件的一种。常用的 LED 发光器件有 2 类:LED 显示器和点阵。

LED 显示器内部由 7 个条形发光二极管和 1 个小圆点发光二极管组成,根据各管的亮暗组合成字符。根据内部发光二极管的接线形式可分为共阴极和共阳极 2 种。共阴极 LED 显示器公共端接地,共阳极 LED 显示器公共端接电源。每段发光二极管需 5~10mA 的驱动电流才能正常发光,一般需加限流电阻控制电流的大小。

2. LED 显示器显示原理

LED 显示器的 a~g 7 段发光二极管,加正电压的发光,加零电压的不发光,不同亮暗发光二极管的组合就能形成不同的字型,这种组合称为字型码。共阳极和共阴极的字型码是不同。

3. 静态显示驱动

静态显示驱动也称直流显示驱动。静态显示驱动是指每个LED显示器的每1个段码都由1个单片机的I/O端口进行驱动，或者使用如BCD码二－十进制译码器译码进行驱动。静态显示驱动的优点是编程简单，显示亮度高；缺点是占用I/O端口多，如驱动5个LED显示器静态显示，需要5×8=40个I/O端口来驱动，要知道1个89S52单片机可用的I/O端口只有32个，实际应用时必须增加译码驱动器进行驱动，这就增加了硬件电路的复杂性。

4. LED显示器驱动电路

要驱动1个4位LED显示器，如图6-2所示常见的有以下几种LED显示器驱动电路。

（1）使用12个I/O端口。

（2）P2.4～P2.7引脚驱动位选，I/O端口直接驱动位选。

（3）P0驱动段选，I/O端口直接驱动段选。

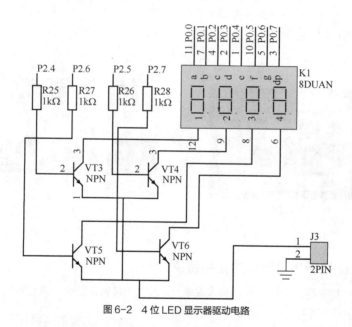

图6-2 4位LED显示器驱动电路

5. 共阴极LED显示器字形码

共阴极LED显示器字形码对应关系如下。

0x3f：0；0x06：1；0x5b：2；0x4f：3；0x66：4；0x6d：5；0x7d：6；0x07：7；0x7f：8；0x6f：9；0x77 A；0x7c：B；0x39：C；0x5e：D；0x79：E；0x71：F；0x00 不显示。

6.2.4 LED显示2种接法

1. LED显示的接法

使用LED显示器时，要注意区分2种不同的接法。为了显示数字或字符，必须对数字或字符进行编码。7段发光二极管加上1个小数点，共计8段，因此为LED显示器提供的编码正好是1个字节。BST实验板用共阴极LED显示器，如图6-3所示。

2. LED显示器显示分析

LED显示器的结构分为共阳极与共阴极，如图6-4所示。

单片机系统扩展 LED 显示器时多用共阳极 LED。共阳极 LED 显示器每段 LED 通过低电平（"0"）驱动发光，要求驱动功率很小；而共阴极 LED 显示器每段 LED 通过高电平（"1"）驱动发光，要求驱动功率较大。通常每段 LED 要串连 1 个数百欧姆的限流电阻。

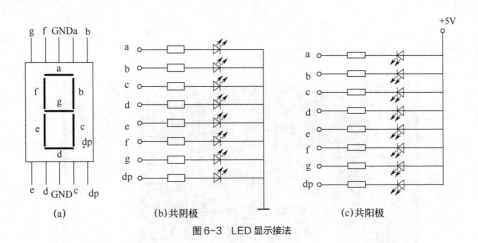

图 6-3 LED 显示接法

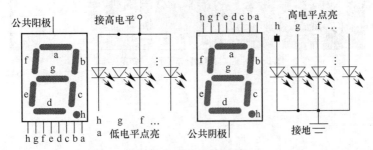

图 6-4 LED 显示器的结构

6.3 项目实施

第 1 个 LED 显示器显示字符"b"的程序如下。

```
/*********************************
* 实验名：LED 显示器静态显示
* 实验效果：第 1 个 LED 显示器显示字符"b"
*********************************/
#include<reg52.h>
#define duan P0
sbit wei1 = P2^4;// 定义第 1 位 LED 显示器
sbit wei2 = P2^5;// 定义第 2 位 LED 显示器
sbit wei3 = P2^6;// 定义第 3 位 LED 显示器
sbit wei4 = P2^7;// 定义第 4 位 LED 显示器

void main()
{
```

```
    wei1 = 1;
    wei2 = 0;
    wei3 = 0;
    wei4 = 0;
    duan = 0x7c;//0111 1100 "b"
    while(1);
}
```

LED 显示器静态显示实验结果实物图如图 6-5 所示。

图 6-5　LED 显示器静态显示实验结果实物图

Chapter 7

项目7
LED显示器动态显示

项目目标

- 了解LED显示器动态显示原理，掌握其编程方法。

建议学时

- 4学时。

知识要点

- 数组、元素、利用数组实现查表。
- 局部变量和全局变量的作用域、可见性。

技能掌握

- 掌握数码管的结构、分类，能用万用表判别数码管的类型。

7.1 项目分析

通过演示 LED 显示器动态显示的操作过程，穿插讲解数组、元素和利用数组实现查表，以及局部变量和全局变量的作用域、可见性等。

7.2 技术准备

7.2.1 LED 显示器动态显示与扫描原理

在熟练掌握 1 位 LED 显示器与单片机的连接，及其上显示 1 位十进制数方法的基础上，进一步掌握用 4 个 1 位 LED 显示器显示 4 位十进制数，及 8 个 1 位 LED 显示器与单片机的连接方法。

4 位及 4 位以上 LED 显示器更适合动态显示，动态显示的连接方式就是将与每一位 LED 显示器的段选线并联在一起，由位选线控制是哪一位数码管有效。动态显示与静态显示相比，虽然具有需要动态刷新的缺点，但同时也解决了静态显示占用 I/O 端口线多的问题，具有占用 I/O 端口线少的优点）。目前已有 4 位一体 LED 显示器内部即为动态显示的连接方式，可方便地选购和使用。

在实际的单片机系统中，往往需要多位显示。动态显示是一种最常见的多位显示方法，应用非常广泛。所有 LED 显示器段选都连接在一起的时候，如何让 LED 显示器显示不一样的数字。动态显示是多个 LED 显示器交替显示，利用人的视觉暂停作用使人看到多个 LED 显示器同时显示的效果。就如电影是由一帧一帧的画面组成，当画面显示速度足够快的时候，看到的就是动态的效果。当 LED 显示器的显示速度足够快的时候，同样也可以看到同时显示的效果。

动态显示的特点是将所有位 LED 显示器的段选线并联在一起，由位选线控制是哪一位 LED 显示器有效。这样就没有必要为每一位 LED 显示器配一个锁存器，从而大大简化了硬件电路。选亮 LED 显示器采用动态扫描显示。所谓动态扫描显示即轮流向各位 LED 显示器送出字形码和相应的位选，利用发光二极管的余辉和人眼视觉暂留作用，使人的感觉好像各位数码管同时都在显示。动态显示的亮度比静态显示要差一些，所以在选择限流电阻时应略小于静态显示电路中的。

7.2.2 4 位 LED 显示器的动态和静态显示连接方式图

4 位 LED 显示器的静态显示连接方式如图 7-1 所示；动态显示连接方式如图 7-2 所示。

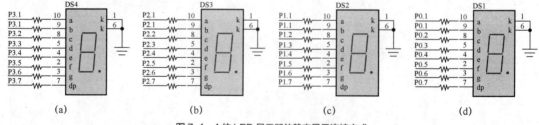

图 7-1 4 位 LED 显示器的静态显示连接方式

7.2.3 编程实验理论准备

本项目的实验是编写一个程序，在 4 位一体 LED 显示器左数第 4 位显示 4，过 1s，在左数第 3 位显示 3，过 1s，在左数第 2 位显示 2，过 1s，在左数第 1 位显示 1s，上述过程不断循环。

将中间的延时时间不断改短，当刷新频率 >50HZ，就感觉不到闪烁了。

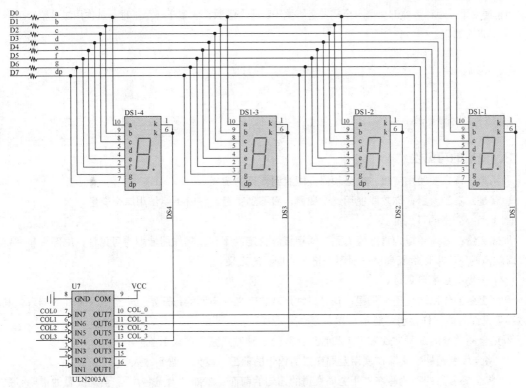

图 7-2　4 位 LED 显示器的动态显示连接方式

1. 变量的作用域

变量有名称、储存单元地址，还有作用范围，也叫有效范围及生命期。变量的作用域是指变量的"有效范围"，具体是指可以使用变量名的程序代码区域。在变量的作用域中，可以合法地引用变量、设置变量的值。在变量的作用域之外，则不能使用变量，否则，将会导致编译错误。可以在整个程序中的所有范围内起作用的变量，称为全局变量。只能在整个程序中的一定的范围内起作用的变量，称为局部变量。

2. 局部变量和全局变量

（1）局部变量。

一对{}括起来的代码范围，属于一个局部作用域。在局部作用域内定义的变量，称为"局部变量"，也称为内部变量。局部作用域可以是一个函数体，也可以是复合语句。在局部作用域内定义的变量，其有效范围从它定义的行开始，一直到该局部作用域结束。离开该局部作用域后，再使用这种变量是非法的。

定义方法如下。

①

```
{
数据类型变量名；
……
}
```

②

```
{
数据类型变量名 = 初值；
……
}
```

（2）全局变量。

在所有函数外部定义的变量具有全局作用域，即该变量在整个工程的所有文件中都是有效的，这种变量就称为全局变量,也称为外部变量。全局变量不属于哪一个函数,它属于一个工程,其作用域是整个工程。

定义方法如下。

①

```
数据类型变量名;
……
```

②

```
数据类型变量名 = 初值;
……
```

3. 变量的可见性

和函数一样,要使用变量,就必须保证能看到变量的定义。

也就是,看到变量,你才能使用这个变量;看不到变量,你就不能使用这个变量。

(1)局部变量的可见性。

局部变量在其作用域内肯定能看到,作用域外肯定看不到,其作用域即为可见性;局部变量在定义的同时就声明了,所以局部变量一般不严格区分变量的定义和声明。

(2)全局变量的可见性。

全局变量的作用域是整个工程;其可见性是从它定义或声明的行开始,一直到源文件的结束,其作用域和可见性可能一样,也可能不一样,即作用域内不一定可见。

看见全局变量的3种方式如下(和看见函数的3种方式完全一致)。

① 将全局变量的定义写在使用全局变量的语句的前面,就如"看到本人"。

② 将全局变量的声明写在使用全局变量的语句的前面,就如"看见名片"。全局变量的声明就是全局变量的"名片",看不到"本人",看到其"名片"也可以。声明形成为

```
extern <类型名><变量名>;
```

全局变量只能在定义的时候初始化,在声明的时候不允许初始化,一次定义,可多次声明。

③ 使用头文件。

4. 关于变量定义的一些建议

全局变量主要用于函数间数据传送,从模块化编程的要求看,函数间数据传送一般要使用形参/实参、返回值来传送,除非万不得已,一般不要用全局变量来传递。总之,要严格控制全局变量的使用,用得越少越好。一般定义变量的时候,没有特殊情况,全部定义为局部变量,使用局部变量的好处如下。

(1)不用考虑本作用域外是否有同名变量。

(2)节约 RAM 空间。

(3)便于移植。

5. 定义一个简单的数据类型

(1)整型无符号 uint 范围 0 ~ 65 535 16 位,定义 uint 为无符号整形语句为

```
#define uint unsigned int
```

(2)字符型无符号 uchar 范围 0 ~ 255 8 位,定义 uchar 为无符号整形语句为

```
#define uchar unsigned char
```

只要定义上面的数据类型后,写程序会简单很多。

6. 数组

数组就是同类型的一批数据的有序集合。由若干个元素组成。必须先定义,后使用。数组的定义为

```
[存储器类型] 数据类型数组名[元素的个数];
```

定义的同时可以对整个数组赋初值;初值的个数不能超过数组的大小;可以不指定元素的个数,则初值的个数就是元素的个数;定义以后只能对单个元素进行赋值;元素的引用用数组名和下标确定。

需要强调，下标从 0 开始。定义数组 uchar sz[10]；得到的是 sz[0] ~ sz[9] 这 10 个元素，sz[10] 是不存在的。

利用数组实现查表的语句为

```
uchar code sz[]
```

7.3 项目实施

1. LED 显示器间隔 1s 的动态显示实验程序

```
/*********************************************
*实验名：LED 显示器动态显示
*实验效果：在 4 位一体 LED 显示器左数第 4 位显示 4，过 1s,
*在左数第 3 位显示 3，过 1s，在左数第 2 位显示 2，过 1s,
*在左数第 1 位显示 1，上述过程不断循环
**********************************************/
#include<reg52.h>
#define duan P0
#define uchar unsigned char
sbit wei1 = P2^4;//定义第 1 位 LED 显示器
sbit wei2 = P2^5;//定义第 2 位 LED 显示器
sbit wei3 = P2^6;//定义第 3 位 LED 显示器
sbit wei4 = P2^7;//定义第 4 位 LED 显示器

uchar code sz[17]={0x3f , 0x06 , 0x5b ,0x4f , 0x66 , 0x6d ,0x7d ,
0x07 , 0x7f , 0x6f ,0x77 , 0x7c , 0x39 , 0x5e , 0x79 , 0x71 , 0x00};

void delay (unsigned int xms)
{
   unsigned int i,j;
   for (i=xms;i>0;i--)                 //i=xms 即延时 xms
      for (j=112;j>0;j--) ;
}

void main()
{
   while (1)
      {
         duan = sz[4];
         wei1 = 0;
         wei2 = 0;
         wei3 = 0;
         wei4 = 1;
         delay (1000);
         duan = sz[3];
         wei1 = 0;
         wei2 = 0;
         wei3 = 1;
         wei4 = 0;
         delay (1000);
         duan = sz[2];
         wei1 = 0;
         wei2 = 1;
         wei3 = 0;
```

```
            wei4 = 0;
            delay (1000);
            duan = sz[1];
            wei1 = 1;
            wei2 = 0;
            wei3 = 0;
            wei4 = 0;
            delay (1000);
        }
}
```

2. LED 显示器间隔 5ms 的动态显示实验程序

```
/*****************************************************
*实验名：LED 显示器动态显示
*实验效果：在 4 位一体 LED 显示器左数第 4 位显示 4，过 5ms，
*在左数第 3 位显示 3，过 5ms，在左数第 2 位显示 2，过 5ms,
*在左数第 1 位显示 1，上述过程不断循环
*注意，每位 LED 显示器之间的延时＜5ms，可看见 4 位 LED 显示器同时点亮
******************************************************/
#include<reg52.h>
#define duan P0
#define uchar unsigned char
sbit wei1 = P2^4;//定义第 1 位 LED 显示器
sbit wei2 = P2^5;//定义第 2 位 LED 显示器
sbit wei3 = P2^6;//定义第 3 位 LED 显示器
sbit wei4 = P2^7;//定义第 4 位 LED 显示器

uchar code sz[17]={0x3f , 0x06 , 0x5b ,0x4f , 0x66 , 0x6d ,0x7d ,
0x07 , 0x7f , 0x6f ,0x77 , 0x7c , 0x39 , 0x5e , 0x79 , 0x71 , 0x00};

void delay (unsigned int xms)
{
    unsigned int i,j;
    for (i=xms;i>0;i--)              //i=xms 即延时 xms
        for (j=112;j>0;j--);
}

void main()
{
    while (1)
        {
            duan = sz[4];
            wei1 = 0;
            wei2 = 0;
            wei3 = 0;
            wei4 = 1;
            delay (5);
            duan = sz[3];
            wei1 = 0;
            wei2 = 0;
            wei3 = 1;
            wei4 = 0;
            delay (5);
            duan = sz[2];
            wei1 = 0;
            wei2 = 1;
```

```
            wei3 = 0;
            wei4 = 0;
            delay(5);
            duan = sz[1];
            wei1 = 1;
            wei2 = 0;
            wei3 = 0;
            wei4 = 0;
            delay(5);
        }
}
```

LED 显示器动态显示实验结果实物图如图 7-3 ~ 图 7-6 所示。

图 7-3　LED 显示器动态显示实验结果实物图 1

图 7-4　LED 显示器动态显示实验结果实物图 2

图 7-5　LED 显示器动态显示实验结果实物图 3

图 7-6　LED 显示器动态显示实验结果实物图 4

此项目附操作视频。

Chapter 8

项目8
独立键盘输入

项目目标

- 掌握键盘输入模块的基本原理及其独立输入程序编制方法。

建议学时

- 4学时。

知识要点

- 各种按键。
- P3引脚。
- IF语句（条件分支语句）。
- 键盘种类。

技能掌握

- 熟悉C语言IF语句的运用，了解单片机的引脚功能。

8.1 项目分析

单片机也需要人机交互,掌握常用的输入和输出设备非常必要。本章选取了最常用的输入模块键盘来演示其典型程序的编制方法。

8.2 技术准备

8.2.1 独立键盘输入理论知识

1. P3 引脚端口第二功能表

P3 引脚就是指 P3.0 ~ P3.7 为双功能端口引脚,对应端口内置上拉电阻具有特定的第二功能,见表 8-1。在不使用第二功能时,这些端口就是普通的通用准双向 I/O 端口引脚。

表 8-1 P3 引脚端口第二功能表

引　　脚	第 二 功 能
P3.0	RxD:串行口接收数据输入端
P3.1	TxD:串行口发送数据输出端
P3.2	INT0:外部中断申请输入端 0
P3.3	INT1:外部中断申请输入端 1
P3.4	T0:外部计数脉冲输入端 0
P3.5	T1:外部计数脉冲输入端 1
P3.6	WR:写外设控制信号输出端
P3.7	RD:读外设控制信号输出端

2. 读端口和读引脚

读端口就是读端口寄存器;读引脚就是读该引脚在端口寄存器中的对应位,通过引用端口寄存器的值,或者引用端口寄存器中的对应位,就可以实现读端口或者读引脚。

3. if 语句

键盘输入程序中经常会用到 if 语句(条件分支语句)。

if 语句可以是复合语句。复合语句就是用 "{ }" 将多条语句组合在一起而形成的一种语句,不需要用 ";" 结束,但它内部的语句仍需要用 ";" 结束。

复合语句格式为

```
{
局部变量定义;
语句 1;
语句 2;
……
语句 n;
}
```

8.2.2 硬件模块工作原理

图 8-1 所示为 4 位按键电路,键按下和释放时,输入信号存在抖动现象,如图 8-2 所示,这是因为按键

利用的是机械触点的闭合、断开作用,由于机械触点具有弹性,在其闭合、断开瞬间均有抖动过程,抖动时间一般在 5～10ms,稳定闭合时间由操作人员的按键动作决定,一般为零点几秒到几秒。为了保证单片机对一次闭合,仅作一次键输入操作,必须在编程时候编写必要的程序代码来去除抖动影响,称键盘软件消抖。

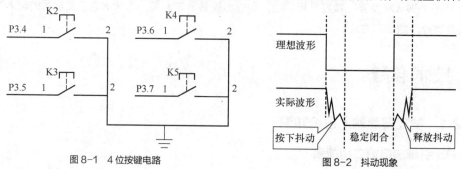

图 8-1　4 位按键电路　　　　　　图 8-2　抖动现象

先判断键是否按下,若按下了延时 10ms,跳过按下抖动期,然后再判断按键是否按下,若是说明按键真的按下了,否则说明是干扰信号,如果键真的按下了,则等待键释放,如果键释放了,延时 10ms,再判断键是否释放,若释放了,说明按键真的释放了,否则说明是干扰信号,如果按键真的释放了,说明一次完整的按键过程完成了。在一次完整的按键后,可以连接该键对应的功能程序段,以实现特定的功能。软件消抖的编程思路框图如图 8-3 所示。

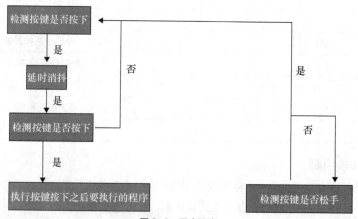

图 8-3　程序思路

8.2.3　认识轻触开关按键

轻触开关是一种电子开关,如图 8-4 所示。使用时,轻轻按开关按钮,就可使开关接通,当松开手时,开关断开。

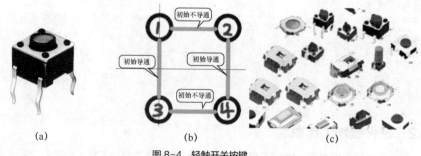

图 8-4　轻触开关按键

8.2.4 键盘的分类

键盘分编码键盘和非编码键盘。键盘上闭合键的识别由专用的硬件编码器实现，并产生键编码号或键值的键盘称为编码键盘，如计算机键盘。而靠软件编程来识别的键盘称为非编码键盘。在单片机组成的各种系统中，用得最多的是非编码键盘，也有用到编码键盘的。非编码键盘分为独立键盘和行列式（又称为矩阵式）键盘。

8.3 项目实施

1. 多按键控制实验程序

```
/**********************************************
* 实验名：独立键盘输入
* 实验效果：按下第 1 个独立按键 K1  2 个红色的 LED 亮
           按下第 2 个独立按键 K1  2 个绿色的 LED 亮
           按下第 3 个独立按键 K1  2 个黄色的 LED 亮
           按下第 4 个独立按键 K1  2 个蓝色的 LED 亮
***********************************************/
#include<reg52.h>

sbit led1 = P1^0;//RED
sbit led2 = P1^1;//RED
sbit led3 = P1^2;//GREEN
sbit led4 = P1^3;//GREEN
sbit led5 = P1^4;//YELLOW
sbit led6 = P1^5;//YELLOW
sbit led7 = P1^6;//BLUE
sbit led8 = P1^7;//BLUE

sbit k1 = P3^4;
sbit k2 = P3^5;
sbit k3 = P3^6;
sbit k4 = P3^7;

void main()
{
   P1 = 0xff;// 熄灭所有 LED
   while(1)
   {
   led1 = k1;
   led2 = k1;
   /*   if(k1 == 0)
        {
            led1 = 0;
            led2 = 0;
        }
        else
        {
            led1 = 1;
            led2 = 1;
        }
   */
   led3 = k2;
   led4 = k2;
   led5 = k3;
```

```
       led6 = k3;
       led7 = k4;
       led8 = k4;
       }
}
```

2. 单按键控制实验程序

```
/*********************************
*实验名：独立键盘输入
*实验效果：按下第1个独立按键K1松开按键后
          第1个led改变状态
*********************************/
#include<reg52.h>

sbit led1 = P1^0;
sbit k1 = P3^4;

void delay(unsigned int xms)
{
  unsigned int i,j;
  for(i=xms;i>0;i--)       //i=xms 即延时 xms
     for(j=112;j>0;j--);
}

void main()
{
  P1 = 0xff;//熄灭所有led
  while(1)
  {
    if(k1 == 0)//判断是否有按下按键的信号
    {
       delay(10);//延时10ms 消抖
       if(k1 == 0)//再次判断按键是否被按下
       {
          while(k1 == 0);//直到按键判断松开
          led1 = ~led1;//松开后执行程序翻转led
       }
    }
  }
}
```

按键实验结果实物图如图8-5所示。

图8-5 按键实验实物图

此项目附操作视频。

Chapter 9

项目9
单片机中断系统

项目目标

- 中断系统是单片机非常重要的组成部分，是为了使单片机能够对外部或内部随机发生的事件实时处理而设置的。中断功能的存在，在很大程度上提高了单片机实时处理能力，也是单片机最重要的功能之一，是学习单片机必须掌握的重要内容。了解中断概念，以INT外部中断为例，详细讲解中断程序的编写方法。

建议学时

- 4学时。

知识要点

- 中断的概念。
- 中断优先级控制。
- 中断的嵌套。

技能掌握

- 学会中断的引发及超过定时器时间范围的延迟函数的编写。

9.1 项目分析

本项目通过定时器中断方式来实现流水灯控制，以及通过计数和定时相结合的方式实现长时间定时。通过中断方式和上一章查询方式的编程差异进行比较，了解并掌握中断的使用。

9.2 技术准备

9.2.1 单片机中断系基本概念

一个高速主机和一个低速外设连接时，效率极低，低速外设工作时无端大量占用 CPU 时间。一个高速主机和多个低速外设连接时，高速主机无法进行多任务并行处理。此时需引入"中断"。

CPU 在处理某一事件 A 时，另一事件 B 发出请求（中断请求）；CPU 暂时中断当前的工作，转去处理事件 B（中断响应和中断服务）；待 CPU 将事件 B 处理完毕后，再回到原来事件 A 被中断的地方继续处理事件 A（中断返回），这一过程称为中断。

引起 CPU 中断的根源，称为中断源。中断源向 CPU 提出中断请求，CPU 暂时中断原来正在处理的事件 A，转去处理事件 B。对事件 B 处理完毕后，再回到原来被中断的地方（即断点），称为中断返回。实现上述中断功能的部件称为中断系统（中断机构）。MC51 单片机中断过程示意图如图 9-1 所示。

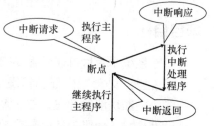

图 9-1 MC51 单片机中断过程示意图

9.2.2 中断传送方式及其特点

早期的计算机系统是不包含中断系统的。后来为了解决快速主机与慢速外设的数据传送问题，引入了中断系统。中断系统具有如下优点。

① 分时操作。CPU 可以分时为多个外设服务，提高了计算机的利用率。

② 实时响应。CPU 能够及时处理应用系统的随机事件，系统的实时性大大增强。

③ 可靠性高。CPU 具有处理设备故障及掉电等突发性事件能力，从而使系统可靠性提高。

在中断传送方式下，平时各自做工作的数据传送的甲，乙双方，一旦甲方要求与乙方进行数据传送，就主动发出信号提出申请，乙方接到申请后若同意传送，安排好当前的工作，再响应，与甲方发生数据传送。传送完毕后，返回，继续做打断前的工作。

中断涉及如下几个环节。

① 中断源（前面提到的甲方）。

② 中断请求（甲方发出信号提出申请）。

③ 开放中断（乙方同意传送）。

④ 保护现场（安排好当前的工作）。

⑤ 中断服务（响应乙方的要求）。

⑥ 恢复现场（传送完成后，回去）。

⑦ 中断返回（继续做打断前的工作）。

中断功能强弱是计算机性能优劣的重要标志。中断传送方式具有如下主要特点。

① 提高 CPU 效率。

② 解决速度矛盾。
③ 实现并行工作。
④ 应付突发事件。

9.2.3 80C51中断系统

1. 中断系统的结构

80C51的中断系统有5个中断源，2个优先级，可实现二级中断嵌套。中断系统的结构如图9-2所示。

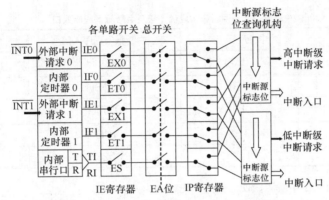

图9-2 中断系统结构

（1）(P3.2) 可由IT0（TCON.0）选择其为低电平有效还是下降沿有效。当CPU检测到P3.2引脚上出现有效的中断信号时，中断标志IE0（TCON.1）置1，向CPU申请中断。

（2）(P3.3) 可由IT1（TCON.2）选择其为低电平有效还是下降沿有效。当CPU检测到P3.3引脚上出现有效的中断信号时，中断标志IE1（TCON.3）置1，向CPU申请中断。

（3）TF0（TCON.5），片内定时/计数器T0溢出中断请求标志。当定时/计数器T0发生溢出时，置位TF0，并向CPU申请中断。

（4）TF1（TCON.7），片内定时/计数器T1溢出中断请求标志。当定时/计数器T1发生溢出时，置位TF1，并向CPU申请中断。

（5）RI（SCON.0）或TI（SCON.1），串口中断请求标志。当串口接收完1帧串行数据时，置位RI，或当串口发送完1帧串行数据时，置位TI，向CPU申请中断。

2. 中断处理过程

（1）中断优先级控制

80C51中断系统有2个中断优先级，即可实现二级中断服务嵌套。80C52中断系统有6个中断源，2个中断优先级，也可实现二级嵌套。STC 2C5A60S2中断系统有10个中断源，2个或4个中断优先级，可实现二级或四级中断服务嵌套。每个中断源的中断优先级都是由中断优先级寄存器IP中的相应位的状态来规定的，见表9-1。IP中断优先级寄存器地址为B8H。

表9-1 中断优先级寄存器IP

位	7	6	5	4	3	2	1	0
功能			PT2	PS	PT1	PX1	PT0	PX0

PX0（IP.0），外部中断 0 优先级设定位。
PT0（IP.1），定时/计数器 T0 优先级设定位。
PX1（IP.2），外部中断 1 优先级设定位。
PT1（IP.3），定时/计数器 T1 优先级设定位。
PS（IP.4），串口优先级设定位。
PT2（IP.5），定时/计数器 T2 优先级设定位。

同一优先级中的中断请求不止一个时，因此有中断优先权排队问题。同一优先级的中断优先权排队，由中断系统硬件确定的自然优先级形成，其排列见表 9-2。

表 9-2 各中断源响应优先级及中断服务程序入口

中 断 源	中 断 标 志	中断服务程序入口	优先级顺序
外部中断 0（$\overline{INT0}$）	IE0	0003H	高
定时/计数器 T0	TF0	000BH	↓
外部中断 1（$\overline{INT1}$）	IE1	0013H	↓
定时/计数器 T1	TF1	001BH	↓
串口	RI 或 TI	0023H	低

设置 51 单片机的 4 个中断源，使他们的优先顺序为

T1，INT1，INT0，T0。

IPH = 0X08，PT1 = 1，IP = 0X40，PX1 = 1。

80C51 中断优先级有如下 3 条原则。

① CPU 同时接收到几个中断请求时，首先响应优先级别最高的中断请求。

② 正在进行的中断过程不能被新的同级或低优先级的中断请求所中断。

③ 正在进行的低优先级中断服务，能被高优先级中断请求所中断。

为了实现上述后 2 条原则，中断系统内部设有 2 个用户不能寻址的优先级状态触发器。其中一个置 1，表示正在响应高优先级的中断，它将阻断后来所有的中断请求；另一个置 1，表示正在响应低优先级中断，它将阻断后来所有的低优先级中断请求。

（2）中断允许控制。

CPU 对中断系统所有中断以及某个中断源的开放和屏蔽是由中断允许寄存器 IE 控制的，见表 9-3。IE 中断允许寄存器的地址为 A8H.

表 9-3 中断允许寄存器 IE

位	7	6	5	4	3	2	1	0
功能	EA			ES	ET1	EX1	ET0	EX0

EX0（IE.0），外部中断 0 允许位。
ET0（IE.1），定时/计数器 T0 中断允许位。
EX1（IE.2），外部中断 0 允许位。
ET1（IE.3），定时/计数器 T1 中断允许位。
ES（IE.4），串口中断允许位。
EA（IE.7），CPU 中断允许（总允许）位。
定时器/计数器控制寄存器 TCON 相应位见表 9-4。

表 9-4 定时器/计数器控制寄存器 TCON

位	D7	D6	D5	D4	D3	D2	D1	D0
功能	TF1	TR1	TF0	TR0	IE1	IT1	IE0	IT0

IT0，IT1，设置外部中断的触发方式。

为 0 时，低电平触发方式。

为 1 时，负跳变触发方式。

IE0，IE1，外部中断标志位。

其他的是定时/计数器的控制。

TF0，TF1，定时器的中断标志。

TR1，TR0，打开相应的定时器。

3．一次中断过程的完整步骤

（1）中断响应有如下 3 个条件

① 中断源有中断请求。

② 此中断源的中断允许位为 1。

③ CPU 开中断（即 EA=1）。

以上 3 条同时满足时，CPU 才有可能响应中断请求。

（2）中断过程的步骤

① 中断请求。中断事件一旦发生，中断源就提交中断请求（将中断标志位置 1），欲请求 CPU 暂时放下目前的工作，转向为该中断作专项服务。

② 中断使能。虽然中断源提交了中断请求，但是，能否得到 CPU 的响应，还要取决于该中断请求能否通过若干关卡送达 CPU（中断使能位等于 1，关卡放行），这些关卡有以下 2 类。

a. 此中断源的中断允许位；

b. 全局中断允许位。

③ 中断响应。如果一路放行，则 CPU 响应该中断请求，记录断点，跳转到中断服务程序。对于 INT 和 TMR 中断，中断响应时中断标志位会被硬件自动清零。

④ 中断处理。对中断源进行有针对性的服务。

⑤ 中断返回。返回到主程序断点处，继续执行主程序。

①，③，⑤由硬件自动完成。②，④是用户编程完成。中断响应条件为①，②同时满足。

4．中断请求标记的置位和清除

（1）外部中断。

下降沿触发方式条件下，在产生中断请求时由硬件置位（置 1）中断请求标记，当 CPU 响应中断时由硬件清除（清 0）。电平触发方式条件下，中断请求标记由外部中断源控制。具体是，当 CPU 检测到 INT 引脚上出现低电平时，中断标志 IE 由硬件置位，INT 引脚上出现高电平时，中断标志 IE 由硬件清除。

① 定时器中断。

计数溢出时，由硬件置位中断请求标记，当 CPU 响应中断时，则由硬件清除。

② 串口中断。

当串口接收完 1 帧数据后请求中断，由硬件置位中断请求标记 RI，RI 必须由软件清除。当串口发送完 1 帧数据后请求中断，由硬件置位中断请求标记 TI，TI 必须由软件清除。

（2）中断源中断。

每个中断源都有如下 3 个位。

① 优先级定义位。每个中断源可以选择中断优先权（Priority，二选一或四选一）。

② 中断使能位。每个中断源均可使能（Enable）或使不能（Disable）。

③ 中断请求标记位。每个中断源均有独立的中断请求标记（Flag）。

a. 中断请求标记的产生。中断事件发生时，由硬件自动产生。

b. 中断请求标记的清除。中断事件被响应时，由硬件自动清除。

（3）全局中断

EA 为全局中断使能位（总允许位），为 0 时，全局中断禁止；为 1 时，全局中断允许。

5. 中断的嵌套

如果多个中断源同时提出了中断请求，先响应高优先级中断源，后响应低优先级中断源。属于相同优先级的中断源，则根据其内部中断查询顺序，先查询的先响应，后查询的后响应。注意，这个查询是硬件自动完成的，程序员并不需要为此书写语句。

如果一个中断源提出了中断请求，已经转去执行其中断服务程序了，期间又有一个中断源提出了中断请求，CPU 的处理原则是，如果新的中断优先级与当前正在处理的中断是同级的，则不予响应，待当前中断服务程序执行完毕后，再响应；如果新的中断优先级比当前正在处理的中断高，则会发生中断嵌套，如图 9-3 所示。

结论为优先级是程序员指定的。在多中断源程序的编写中，程序员必须认真考虑优先级问题，否则中断系统会运行不正常，甚至导致危险的发生。

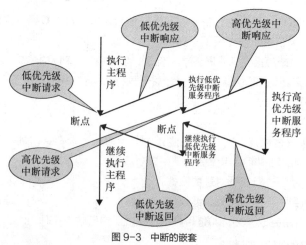

图 9-3　中断的嵌套

9.2.4　中断服务函数

1. 中断服务函数的写法

中断服务函数的写法为

void 函数名 () interrupt 中断编号
{
　；
}

定义中断函数的一般形为

viod 函数名 () interrupt n

需要注意以下 2 点。

（1）interrupt 必须要加，表示定义成中断服务函数。

（2）后面 n 是中断号，中断号是编译器识别不同中断的唯一编号。

2. 中断服务函数和普通函数的异同

（1）相同点是，函数的形式非常类似，中断响应过程和普通函数调用过程也非常相似。

（2）不同点有如下几方面。

① 中断服务函数不需要声明，普通函数一般需要声明。

② 普通函数的执行是可预测的；中断服务函数的执行是不可预测的。

③ 普通函数的跳转是软件（函数调用语句）完成的；中断服务的跳转（中断响应）是由硬件完成的，只要发生了中断事件，并且中断被允许，硬件自动完成中断服务的跳转（中断响应）。

④ 普通函数通过函数名找到被调用函数；中断服务函数通过中断号找到中断服务函数。由此可知，中断服务函数中的函数名其实并没有什么作用。

3. 使用中断服务函数应遵循的规则

使用中断服务函数时应遵循以下规则。

（1）中断服务函数不能进行参数转递。

（2）在任何情况下，都不能直接调用中断服务函数。

9.3 项目实施

9.3.1 外部中断低电平触发

外部中断低电平触发实验程序如下。

```
/******************************************
*实验名：外部中断实验（低电平触发）
*实验效果：给 P3.2 引脚低电平触发中断，使 LED
*         在触发前后 4 位电平状态调换
*******************************************/
#include<reg52.h>

void delay(unsigned int xms)
{
   unsigned int i,j;
   for(i=xms;i>0;i--)      //i=xms 即延时 xms
      for(j=112;j>0;j--);
}

void main()
{
   P1 = 0x0f;// 0000 1111 下面 4 个灯亮，上面 4 个灯灭
   EX0 = 1;//INT0 中断允许
   EA = 1;// 全局中断打开
   IT0 = 0;// 触发方式为低电平触发
   while(1);
}

void low()interrupt 0
{
   P1 = ~ P1;
   delay(200);
}
```

注意，进行本项目程序，先用杜邦线将单片机 P3.2 引脚与最右下角端口连接起来，如图 9-4 所示。

图 9-4 外部中断低电平触发实验实物接线图

9.3.2 外部中断下降沿触发

外部中断下降沿触发实验程序如下。

```c
/*****************************************
* 实验名：外部中断实验（下降沿触发）
* 实验效果：按下按键触发 P3.3 下降沿，使 LED
*          在触发前后 4 位电平状态调换
*****************************************/
#include<reg52.h>

void main()
{
    P1 = 0x0f;// 0000 1111 下面 4 个灯亮，上面 4 个灯灭
    EX1 = 1;//INT1 中断允许
    EA = 1;// 全局中断打开
    IT1 = 1;// 触发方式为下降沿触发
    while(1);
}

void jump_low() interrupt 2
{
    P1 = ~ P1;
}
```

注意，进行本项目程序，先用杜邦线将单片机 P 3.3 与 P 3.6 引脚端口连接起来，如图 9-5 所示。

图 9-5 外部中断下降沿触发实验实物图

此项目附操作视频。

Chapter 10

项目10
单片机定时器

项目目标

- 通过定时器/计数器实现流水灯控制。

建议学时

- 4学时。

知识要点

- 定时器的结构。
- TMOD和TCON。
- 定时/计数器工作方式。
- 定时/计数器编程步骤。

技能掌握

- 学会定时器的设置、计数器的设置,以及掌握采用查询方式使用定时器。

10.1 项目分析

前面的流水灯的时间控制通过空循环语句来实现，定时不是很精确。本章通过用定时器来控制流水灯任务，可以实现精确的时间控制。这就需要了解定时器的使用。定时器和计数器实质功能相同，本章利用 LED 二进制计数任务来掌握计数器的使用。

10.2 技术准备

10.2.1 单片机定时基础

前面讲过"用若干次空循环实现延时"的生活中的例子均为附件延时。软件延时的缺点是，延时过程中，CPU 时间被占用，无法进行其他任务，导致系统效率降低。延时时间越长，该缺点便越明显，因此软件延时只适用于短暂延时，或简单项目。

定时/计数器的使用，实现了单片机对时间更有效地控制。单片机中有多个小闹钟，可以实现延时，这些小闹钟就是"定时器"。本讲只讲定时/计数器 T0。每个定时/计数器既可以实现定时功能，也可以实现计数功能，本讲只讲定时功能。

10.2.2 定时/计数器 T0 的工作原理

在定时方式 1 下，定时/计数器 T0 的核心是一个 16 位宽的，由计数脉冲触发的，按递增规律（即累加方式）工作的循环累加计数器（TH0+TL0）。从预先设定的初始值开始，每来 1 个计数脉冲就加 1，当加到计数器为全 1 时，再输入 1 个脉冲，就会发生溢出现象，计数器回零，同时产生溢出中断请求信号（TF0 置 1）。如果定时/计数器工作于定时模式，则表示定时时间已到。

1. 80C51 的定时/计数器

单片机实现定时功能，比较方便的办法是利用单片机内部的定时/计数器。80C51 实现定时功能可以采用下面 3 种方法。

（1）软件定时。软件定时不占用硬件资源，但占用了 CPU 时间，降低了 CPU 的利用率。

（2）采用时基电路定时。例如采用 555 电路，外接必要的元器件（电阻和电容），即可构成硬件定时电路。但在硬件连接好以后，定时值与定时范围不能由软件进行控制和修改，即不可编程。

（3）采用可编程芯片定时。这种定时芯片的定时值及定时范围很容易用软件来确定和修改，此种芯片定时功能强，使用灵活。在单片机的定时/计数器不够用时，可以考虑进行扩展。

2. 定时器溢出

任何一个计数范围有限的系统，均存在溢出现象，如图 10-1 所示。系统的可表达数的个数，称为模。定时器溢出时会导致定时器溢出中断请求，和该中断是否使能无关。

10.2.3 定时/计数器的结构及工作原理

1. 定时/计数器的结构

定时/计数器的实质是加 1 计数器（16 位），由高 8 位和低 8 位 2 个寄存器组成。TMOD 是定时/计数器的工作方式寄存器，确定工作方式和功能；TCON 是控制寄存器，控制 T0，T1 的启动和停止及设置溢出标志，其结构原理图如图 10-2 所示。

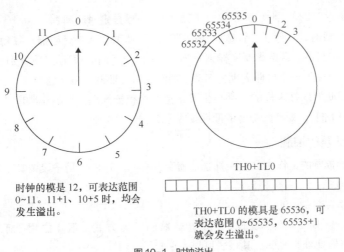

图 10-1 时钟溢出

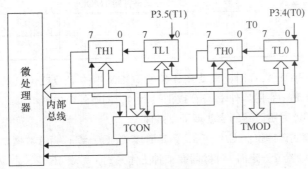

图 10-2 定时/计数器结构原理图

2. 定时/计数器的工作原理

定时/计数器工作原理图如图 10-3 所示。加 1 计数器输入的计数脉冲有 2 个来源，一个是由系统的时钟振荡器输出脉冲经 12 分频后送来；另一个是 T0 或 T1 引脚输入的外部脉冲源。每来 1 个脉冲，计数器加 1，当加到计数器为全 1 时，再输入 1 个脉冲就使计数器回零，且计数器的溢出使 TCON 中 TF0 或 TF1 置 1，向 CPU 发出中断请求（定时/计数器中断允许时）。如果定时/计数器工作于定时模式，则表示定时时间已到；如果工作于计数模式，则表示计数值已满。可见，由溢出时计数器的值减去计数初值才是加 1 计数器的计数值。

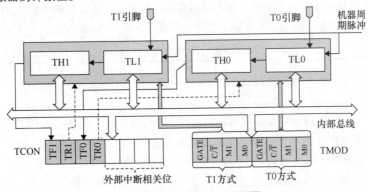

图 10-3 定时/计数器工作原理图

设置为定时器模式时，加 1 计数器是对内部机器周期计数（1 个机器周期等于 12 个振荡周期，即计数频率为晶振频率的 1/12）。计数值 N 乘以机器周期 T_{cy} 就是定时时间 t。

设置为计数器模式时，外部事件计数脉冲由 T0 或 T1 引脚输入到计数器。在每个机器周期的 S5P2 期间采样 T0、T1 引脚电平。当某周期采样到一高电平输入，而下一周期又采样到一低电平时，则计数器加 1，更新的计数值在下一个机器周期的 S3P1 期间装入计数器。由于检测 1 个从 1 到 0 的下降沿需要 2 个机器周期，因此要求被采样的电平至少要维持 1 个机器周期。当晶振频率为 12MHz 时，最高计数频率不超过（1/2）MHz，即计数脉冲的周期要大于 2μs。

3. 定时 / 计数器的控制

80C51 定时 / 计数器的工作由 2 个特殊功能寄存器控制。TMOD 用于设置其工作方式；TCON 用于控制其启动和中断申请。

（1）工作方式寄存器 TMOD。

工作方式寄存器 TMOD 用于设置定时 / 计数器的工作方式，低 4 位用于 T0，高 4 位用于 T1。TMOD 工作方式寄存器地址为 89H。其格式见表 10-1。

表 10-1　工作方式寄存器 TMOD

位	7	6	5	4	3	2	1	0
功能	GATE	C/$\overline{T}$	M1	M0	GATE	C/$\overline{T}$	M1	M0

GATE 为门控位。GATE = 0 时，只要用软件使 TCON 中的 TR0 或 TR1 为 1，就可以启动定时 / 计数器工作；GATA = 1 时，要用软件使 TR0 或 TR1 为 1，同时外部中断引脚或也为高电平时，才能启动定时 / 计数器工作。即此时定时器的启动多了 1 条件。

C/T 为定时 / 计数模式选择位。C/T = 0，为定时模式；C/T =1，为计数模式。

M1M0 为工作方式设置位。定时 / 计数器有 4 种工作方式，见表 10-2，由 M1M0 进行设置。

表 10-2　定时 / 计数器工作方式设置

M1M0	工作方式	说　明
00	方式 0	13 位定时 / 计数器
01	方式 1	16 位定时 / 计数器
10	方式 2	8 位自动重装定时 / 计数器
11	方式 3	T0 分成 2 个独立的 8 位定时 / 计数器；T1 此方式停止计数

（2）控制寄存器 TCON

TCON 的低 4 位用于控制外部中断，已在前面介绍。TCON 的高 4 位用于控制定时 / 计数器的启动和中断申请。TCON 控制寄存器地址为 88H。其格式见表 10-3。

表 10-3　控制寄存器 TCON

位	7	6	5	4	3	2	1	0
功能	TF1	TR1	TF0	TR0	…	…	…	…

TF1（TCON.7）为 T1 溢出中断请求标志位。T1 计数溢出时，由硬件自动置 TF1 为 1。CPU 响应中断后，TF1 由硬件自动清 0。T1 工作时，CPU 可随时查询 TF1 的状态。所以，TF1 可用作查询测试的标志。TF1 也可以用软件置 1 或清 0，同硬件置 1 或清 0 的效果一样。

TR1（TCON.6）为 T1 运行控制位。TR1 置 1 时，T1 开始工作；TR1 清 0 时，T1 停止工作。TR1 由软件置 1 或清 0。所以，用软件可控制定时 / 计数器的启动与停止。

TF0（TCON.5）为 T0 溢出中断请求标志位，其功能与 TF1 同。

TR0（TCON.4）为 T0 运行控制位，其功能与 TR1 类同。

10.2.4 定时/计数器的工作方式

1. 方式 0

方式 0 如图 10-4 所示，为 13 位计数，由 TL0 的低 5 位（高 3 位未用）和 TH0 的 8 位组成。TL0 的低 5 位溢出时向 TH0 进位。TH0 溢出时，置位 TCON 中的 TF0 标志，向 CPU 发出中断请求。

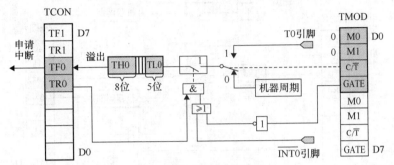

图 10-4 定时/计数器的工作方式 0

2. 方式 1

方式 1 如图 10-5 所示，其计数位数是 16 位，由 TL0 作为低 8 位、TH0 作为高 8 位，组成了 16 位加 1 计数器。

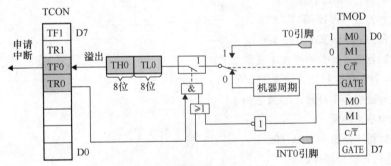

图 10-5 定时/计数器的工作方式 1

计数个数与计数初值的关系为

$$X=2^{16}-N$$

3. 方式 2

方式 2 如图 10-6 所示，为自动重装初值的 8 位计数方式。

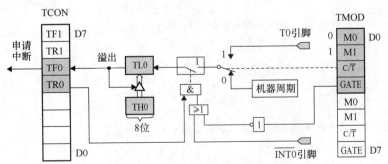

图 10-6 定时/计数器的工作方式 2

计数个数与计数初值的关系为

$$X=2^8-N$$

4．方式3

方式3如图10-7所示，只适用于定时/计数器T0，定时/计数器T1处于方式3时相当于TR1=0，停止计数。

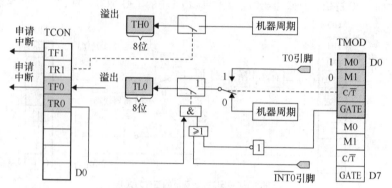

图10-7　定时/计数器的工作方式3

工作方式3时，T0被分成2个独立的8位计数器TL0和TH0。

5．定时器的操作

（1）定时器操作步骤如下。

① 选择工作方式（设置M1，M0）。

② 选择控制方式（设置GATE）。

③ 选择定时器还是计数器模式（设置C/T）。

④ 给定时/计数器赋初值（设置THx和TLx）。

⑤ 开启定时器中断（设置ET0或ET1）。

⑥ 开启总中断（设置EA）。

⑦ 打开计数器（设置TR1或TR0）。

（2）配置定时器值的程序如下。

```
/***************************************************************
* 函数名：TimerConfiguration()
* 函数功能：配置定时器值
* 输入：无
* 输出：无
***************************************************************/
void TimerConfiguration()
{
    TMOD = 0x01;            // 定时器T0选择工作方式1
    TH0 = 0x3C;             // 设置初始值
    TL0 = 0x0B0;
    EA = 1;                 // 打开总中断
    ET0 = 1;                // 打开定时器T0中断
    TR0 = 1;                // 启动定时器T0
}
```

（3）定时器T0初值计算（为方便计算，本讲晶振改用12MHz）。

定时器T0初值计算式为

初值 =65 536- 计数

其中，计数值 = 定时时间 /（振荡周期 ×12）

定时时间 = 振荡周期 ×12×（65 536- 初值）

最大定时时间 + 振荡周期 ×12×65 536=65.536ms

试用 51 单片机定时器初值计算器，快速计算 16 位定时器的初值。

10.3 项目实施

1. 第 1 个 LED 以定时器 T0 设定的时间周期闪烁实验程序

```
/*****************************************
*实验名：定时器实验
*实验效果：第 1 个 LED 以定时器 T0 设定的时间周期闪烁
*****************************************/
#include<reg52.h>

sbit led = P1^0;

void timer0_init()
{
    TMOD = 0x01;    // 定时器 T0 选择工作方式 1
    TH0 = 0x00;             // 设置初始值
    TL0 = 0x00;
    EA = 1;                 // 打开总中断
    ET0 = 1;                // 打开定时器 T0 中断
    TR0 = 1;                // 启动定时器 T0
}

void main()
{
    led = 1;
    timer0_init();
    while (1);
}

void timer0() interrupt 1
{
    TH0 = 0x00;     // 设置初始值
    TL0 = 0x00;
    led = ~ led;
}
```

2. 第 1 个 LED 以 500ms/ 次的精确频率闪烁实验程序

```
/*****************************************
*实验名：定时器实验
*实验效果：第 1 个 LED 以 500ms/ 次的精确频率闪烁
*****************************************/
#include<reg52.h>

sbit led = P1^0;
int i = 0;
```

```c
void timer1_init()
{
    TMOD = 0x10; // 定时器T0选择工作方式1
    TH1 = 0x4C;          // 设置初始值，定时50ms
    TL1 = 0x00;
    EA = 1;                      // 打开总中断
ET1 = 1;                         // 打开定时器T0中断
    TR1 = 1;                     // 启动定时器T0
}

void main()
{
    led = 1;
    timer1_init();// 定时器T1的初始化
    while(1)
    {
      if(i==10)
      {
        led = ~led;
        i = 0; // 注意i需设置为零
      }
    }


}

void timer1() interrupt 3
{
    TH1 = 0x4C;         // 设置初始值，定时50ms
    TL1 = 0x00;
  i++;
}
```

此项目附操作视频及代码资料。

Chapter 11

项目11
串口通信

项目目标

- 了解计算机串行通信基础知识,掌握51单片机的串口编程。

建议学时

- 4学时。

知识要点

- 串口的4种工作方式。
- 双机串行通信的软件编程。
- 串口的4种工作方式。
- 双机串行通信的软件编程。

技能掌握

- 学会通信程序编写。

11.1 项目分析

单片机的串口通信工作方式有 4 种，需要设置更多的寄存器。前面学习过的定时器与中断是单片机通信的基础。本项目的任务依然是流水灯，但需要通过通信方式，将甲机的控制信号发送到乙机实现。

11.2 技术准备

11.2.1 串口通信理论知识

1. 计算机通信的概念

通信是指通过某种媒体将信息从一地传送到另一地。古代飞鸽传书和今天电话、手机，都是人与人之间的通信工具。计算机通信是将计算机技术和通信技术相结合，完成计算机与外部设备或计算机与计算机之间的信息交换。计算机与计算机之间的通信分下面 3 种情况。

① PC 机与 PC 机通信。
② PC 机与单片机通信（本项目只讲这一种）。
③ 单片机与单片机通信。

注意，本项目所提及的计算机包括单片计算机，即单片机。

2. 计算机通信的意义

计算机通信的出现，大大拓展了计算机的应用范围。PC 机与单片机通信，可以实现远程测控，组成计算机网络。

11.2.2 计算机通信的分类

计算机通信是将计算机技术和通信技术相结合，完成计算机与外部设备或计算机与计算机之间的信息交换。计算机通信的分类如图 11-1 所示，其中有线通信有并行通信和串行通信 2 种方式。在多微机系统以及现代测控系统中，信息的交换多采用串行通信方式。

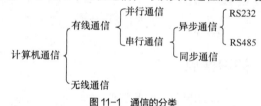

图 11-1 通信的分类

11.2.3 串行通信与并行通信

计算机有线通信可以分为两大类，即并行通信与串行通信。并行通信通常是将数据字节的各位用多条数据线同时进行传送，如图 11-2 所示。串行通信是将数据字节分成一位一位的形式在 1 条传输线上逐个地传送，如图 11-3 所示。

并行通信的特点为，多位数据同时传送，控制简单、传输速度快；由于传输线较多，长距离传送时成本高，且接收方的各位同时接收存在困难。

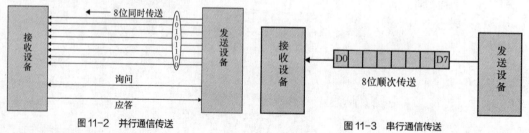

图 11-2 并行通信传送　　　　　　　图 11-3 串行通信传送

串行通信的特点为，数据字节一位一位在一条传输线上逐个传送，传输线少，长距离传送时成本低,且可以利用电话网等现成的设备,但数据的传送控制比并行通信复杂。

1. 串行通信的基本概念

首先介绍异步通信与同步通信。

(1)异步通信。

① 时钟。

收、发设备使用各自时钟,示意图如图 11-4 所示。

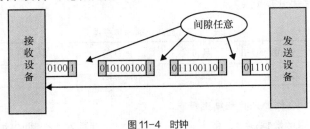

图 11-4 时钟

② 数据格式。

以字符(构成的帧)为单位,字符间是异步的,字符内各位是同步的,其数据格式如图 11-5 所示。

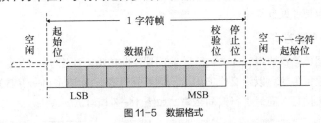

图 11-5 数据格式

(2)同步通信。

发送方时钟与接收方时钟同步,既保持位同步,也保持字符同步。

① 同步方法。

同步方法分为时序同步和位同步 2 种,字符同步方法示意图如图 11-6(a)所示,位同步方法示意图如图 11-6(b)所示。

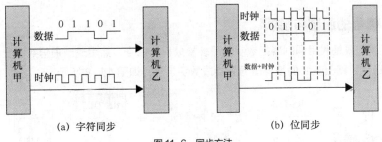

(a) 字符同步　　　　　　　　(b) 位同步

图 11-6 同步方法

② 同步格式。

a. 面向字符的同步格式见表 11-1。

表 11-1 面向字符的同步格式

SYN	SYN	SOH	标题	STX	数据块	ETB/ETX	块校验

(a)同步字符 SYN(16H)。

(b)序始字符 SOH(01H),表示标题的开始。

（c）标题包含源地址、目标地址和路由指示等信息。
（d）文始字符 STX（02H）。
（e）数据块是传送的正文内容，由多个字符组成。
（f）组终字符 ETB（17H）或文终字符 ETX（03H）。
（g）校验码。

例如：IBM 的二进制同步规程 BSC。

b. 面向位的同步格式见表 11-2。

表 11-2 面向位的同步格式

8位	8位	8位	≥0位	16位	8位
01111110	地址场	控制场	信息场	检验场	01111110

（a）用序列 01111110 作为开始和结束标志。
（b）发送方在其发送的数据流中，每出现 5 个连续的 1，就插入 1 个附加的 0；接收方则每检测到 5 个连续的 1，且其后有 1 个 0 时，就删除该 0。

例如：ISO 的高级数据链路控制规程 HDLC 和 IBM 的同步数据链路控制规程 SDLC。

传输效率较高，但硬件设备复杂。

2. 串行通信的传输方向

（1）单工：指数据传输仅能沿 1 个方向，不能实现反向传输，如图 11-7（a）所示。
（2）半双工：指数据传输可以沿 2 个方向，但需要分时进行，如图 11-7（b）所示。
（3）全双工：指数据可以同时进行双向传输，如图 11-7（c）所示。

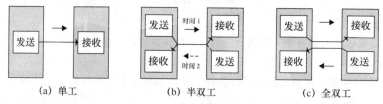

图 11-7 串行通信的传输方向

11.2.4 信号的调制与解调

简单地讲，调制器是利用调制器把数字信号转换成模拟信号，然后送到通信线路上去；而解调器必通过解调器把从通信线路上收到的模拟信号转换成数字信号，如图 11-8 所示。

图 11-8 信号传送

11.2.5 串行通信的错误校验

1. 奇偶校验

在发送数据时，数据位的最后 1 位为奇偶校验位（1 或 0）。奇校验时，数据中"1"的个数与校验位"1"的个数之和应为奇数;偶校验时,数据中"1"的个数与校验位"1"的个数之和应为偶数。接收字符时,对"1"的个数进行校验,若发现不一致,则说明传输数据过程中出现了差错。

2. 代码和校验

代码和校验是发送方将所发数据块求和(或各字节异或),产生1个字节的校验字符(校验和)附加到数据块末尾。接收方接收数据,同时对数据块(除校验字节外)求和(或各字节异或),将所得的结果与发送方的"校验和"进行比较,相符,则无差错;否则即认为传送过程中出现了差错。

3. 循环冗余校验

循环冗余校验通过某种数学运算,实现有效信息与校验位之间的循环校验,常用于对磁盘信息的传输、存储区的完整性校验等。这种校验方法纠错能力强,广泛应用于同步通信中。

11.2.6 传输速率及其与传输距离的关系

传输速率分为工种,即比特率和波特率。
比特率:每秒传输二进制代码的位数。
波特率:每秒调制信号变化的次数,即每秒发送的位数,单位是波特(Baud)。
只有基带传输的比特率和波特率是相同的,二者在其他情况下则不相同。
传输距离与传输速率的关系是,传输距离随传输速率的增加而减小。

11.2.7 串口结构

AT89C51单片机串口的内部结构如图11-9所示,该结构中有2个物理上独立的接收、发送缓冲器SBUF属于特殊功能寄存器,可同时发送、接收数据。它们占用同一地址(99H);接收器是双缓冲结构;发送缓冲器,且因为发送时CPU是主动的,不会产生重叠错误。

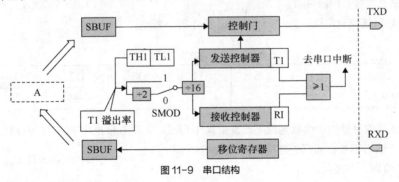

图11-9 串口结构

11.2.8 串行通信的数据结构

串行通信示意图如图11-10所示。在串行通信中,由于异步通信不要求收发双方时钟的严格一致,所以实现容易,设备开销较小,但每个字符要附加2～3位用于起、止位,各帧之间还有间隔,因此传输效率不高。串口的工作方式见表11-3。工作方式寄存器SCON各位功能见表11-4,工作方式寄存器PCONM各位功能见表11-5。

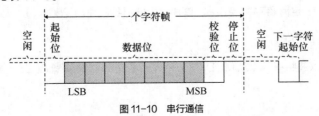

图11-10 串行通信

表 11-3 串口的工作方式

SM0	SM1	方式	说明	波特率
0	0	0	移位寄存器	$f_{osc}/12$
0	1	1	10 位异步收发器（8 位数据）	可变
1	0	2	11 位异步收发器（9 位数据）	$f_{osc}/64$ 或 $f_{osc}/32$
1	1	3	11 位异步收发器（9 位数据）	可变

SM2：多级通信控制位。因为多级通信是在工作方式 2 和工作方式 3 下进行的，因此 SM2 主要用在工作方式 2 和工作方式 3。

当 SM2=0 时，不论接收的第 9 位是 0 还是 1，都接收数据，产生中断。

当 SM2=1 时，只有在接收到的第 9 位为 1 时，才接收数据，并产生中断；而如果接收到的第 9 位为 0 时，则将接收到的数据丢弃，不产生中断。

表 11-4 工作方式寄存器 SCON

位	D7	D6	D5	D4	D3	D2	D1	D0
功能	SM0	SM1	SM2	REN	TB8	RB8	TI	RI

RI：接收中断标志位。接收结束时，会由硬件置 1，向 CPU 发出中断请求。要由软件复位。

TI：发送中断标志位。发送结束时，会由硬件置 1，向 CPU 发出中断请求。要由软件复位。

TB8：用来存放发送的第 9 位。

RB8：用来存放接收的第 9 位。

REN：是串行接收允许位。为 0 时，允许；为 1 时，禁止。

表 11-5 工作方式寄存器 PCON

位	D7	D6	D5	D4	D3	D2	D1	D0
功能	SMOD	—	—	—	—	—	—	—

PCON 是没有位寻址的，也就是说不能直接操作 SMOD，要直接操作 PCON 寄存器。

SMOD：是波特率是否加倍的选择位。

当 SMOD=0 时，不加倍。

当 SMOD=1 时，加倍。

此处，波特率是串口每秒发送的位数，比如，2 400 的波特率就是每秒发送 2 400 位。串口用定时器 T1 作为波特率发生器，这时定时器软件设置在工作方式 2（可自动重装初值）。常用波特率与定时器 1 的参数关系见表 11-6。

因为波特率 =（2SMOD/32）× 定时器 T1 的溢出率

定时器 T1 的溢出率 = 单片机内部时钟频率/（256-X）

式中，X 是定时器的初值。

所以波特率 =（2SMOD/32）× 单片机内部时钟频率 /（256-X）

表 11-6　常用波特率与定时器 T1 的参数关系

串口工作方式及波特率 /kbit/s		f_{osc}/MHz	SMOD	定时器 T1		
				C/$\overline{T}$	工作方式	初值
方式1，3	62.5	12	1	0	2	FFH
	19.2	11.059 2	1	0	2	FDH
	9.6	11.059 2	0	0	2	FDH
	4.8	11.059 2	0	0	2	FAH
	2.4	11.059 2	0	0	2	F4H
	1.2	11.059 2	0	0	2	E8H

串口的操作步骤如下。

(1) 设置波特率。

① 设置定时器 T1 为工作方式 2（设置 TMOD 寄存器）。

② 给计数器赋初值（工作方式 2 会自动重装）。

(2) 设置串口工作方式。

① 设置 SCON（如果允许）。

② 如果使用中断方式，那么打开相应的中断和总中断。

③ 打开定时器 T1，开始产生波特率。

④ 设置 TRx。

设置串口的程序如下。

```
/***********************************************
* 函数名:UsartConfiguration()
* 函数功能:设置串口
* 输入:无
* 输出     :无
***********************************************/
void UsartConfiguration()
{
    SCON=0X50;              // 设置为工作方式 1
    TMOD=0X20;              // 设置计数器工作方式 2
    PCON=0X80;              //SMOD=1，波特率加倍
    TH1=0XF3;               // 计数器初始值设置，注意波特率是 4 800bit/s
    TL1=0XF3;
    ES=1;                   // 打开接收中断
    EA=1;                   // 打开总中断
    TR1=1;                  // 打开计数器
}
```

11.3　项目实施

串口通信实验程序如下。

```
/*********************************************
*实验名：串口通信实验
*实验效果：配合串口助手使用，将 PC 机发出的数据传送到单片机中，
*          单片机将接收到的数据发送到 PC 机，可以在串口助手上显示
*波特率：9 600bit/s
```

```c
*********************************************************/
#include<reg52.h>

void UsartConfiguration()
{
    SCON=0X50;              // 设置为工作方式1
    TMOD=0X20;              // 设置计数器工作方式2
    PCON=0X00;              //SMOD=0,32分频
    TH1=0Xfd;               // 计数器初始值设置,注意波特率是9 600bit/s
    TL1=0Xfd;
    ES=1;                   // 打开接收中断
    EA=1;                   // 打开总中断
    TR1=1;                  // 打开计数器
}

void main()
{
    UsartConfiguration();
    while(1);
}

void uart()interrupt 4
{
    unsigned char date;

    date = SBUF;// 取出接收到的数据
    RI = 0;// 清除接收中断标志位
    SBUF = date;// 将接收到的数据放回发送缓存器
    while(!TI);// 等待发送数据完成
    TI = 0;// 清除发送中断标志位
}
```

串口调试助手界面如图11-11所示。

图11-11 串口调试助手界面

此项目附操作视频及代码资料。

Chapter 12

项目12
综合实验：秒表

项目目标
- 结合前面所有所学的基础实验知识，设计一个秒表。

建议学时
- 8学时。

知识要点
- 所有基础项目所学内容。

技能掌握
- 掌握所有基础项目所学内容。

12.1 项目分析

电子时钟程序由显示模块、校时模块和时钟运算模块三大部分组成,其中校时模块和时钟运算模块要对时、分、秒的数值进行操作,并且秒计数值累加到 60 时,要自己清零,并向分进 1;分计数值累加到 60 时,要自己清零,并向时进 1;时计数值累加到 24 时,要清零。这样,才能循环记时。秒表程序也由显示模块、动 / 暂停复位模块和时钟运算模块组成。其中校时模块和时钟模块要对(1/100)秒、(1/10)秒、秒、分的数值进行操作,并且(1/100)秒累加到 10 时,要自己清零,并向(1/10)秒进 1;(1/10)秒累加到 10 时,要自己清零,并向秒进 1。用按键决定是进入时钟程序还是秒表程序。

12.2 技术准备

熟悉前面所学的基础项目知识,把所有知识结合起来。秒表实验结果实物图如图 12-1 所示,硬件使用了 LED 显示器和按键等。

图 12-1 秒表实验结果实物图

12.3 项目实施

秒表实验程序如下。

```
/*******************BST-M51 实验开发板例程 *********************************
 *   平台:BST-M51 + Keil U4 + STC89C51
 *   名称:秒表
 *   日期:2015-6
 *   晶振:11.059 2MHz
 *   功能说明:1. 按下按键 K4,秒表开始 / 暂停计时
 *            2. 按下按键 K5,秒表清零计时,并等待开始计时
 *            3. 在等待计时过程中,LED D1-D8 闪烁,表示在等待计时
 *            4. 计时过程中,每计时 1 秒,蜂鸣器响一声,并在 LED 显示器上显示
 *            5. LED 显示器上前 2 位显示分,后 2 位显示秒,中间用小数点分隔
 *            6. 计时最大长度为 1 小时,59 分 59 秒后,自动清零计时
 **************************************************************************/
```

```c
// 注意晶振需为11.059 2MHz。
// 若为其他数值晶振，请改变TH0与TL0参数值，否则计时会有很大误差 。

#include<reg51.h>

#define uchar unsigned char

#define dula P0              // 段选信号的锁存器控制
#define wela P2              // 位选信号的锁存器控制，这里只用到P2.4～P2.7引脚

sbit k4 = P3^6;// 红色按键K4
sbit k5 = P3^7;// 红色按键K5
sbit beep = P2^3;// 蜂鸣器

bit stop = 1;
unsigned char j,k,num0,num1,num2,num3,sec,min,count=0;
unsigned char time_count;
unsigned char code weitable[]={0x8f,0x4f,0x2f,0x1f};
//LED显示器各位的码表
unsigned char code table[]={0x3f,0x06,0x5b,0x4f,0x66,0x6d,0x7d,
                            0x07,0x7f,0x6f,0x77,0x7c,0x39,0x5e,0x79,0x71};
//LED显示器各位的码表（带上小点）
unsigned char code table1[]={0xbf,0x86,0xdb,0xcf,0xe6,0xed,0xfd,
                             0x87,0xff,0xef};
void delay(unsigned char i)
{
   for(j=i;j>0;j--)
     for(k=125;k>0;k--);
}

void display1(uchar wei,uchar shu)// 在任意一位显示任意数字
{
   wei=wei-1;
   wela|=0xf0;// 给P2.4～P2.7引脚置1
   if(wei == 2)
    P0=table1[shu];
   else
    P0=table[shu];
   wela=wela&weitable[wei];// 给P2引脚需要显示的那一位置1,其他清0
   delay(5);
}
void display(unsigned char a,unsigned char b,unsigned char c,unsigned char d)
{                              //1次显示4个数字,且每位显示数字可自定
display1(4,a);
   display1(3,b);
   display1(2,c);
   display1(1,d);
}

void start_timer()
{
   ET0=1;                    // 开定时器T0中断
   EA=1;                     // 开总中断
   TR0=1;                    // 打开定时器
}
```

```c
void stop_timer()
{
    ET0=0;                    // 关定时器T0中断
    EA=0;                     // 关总中断
    TR0=0;                    // 关闭定时器
}

void main()
{
    TMOD=0x01;                // 模式设置，00000001，可见采用的是定时器T0，工作与模式1（M1=0，M0=1）
    TH0=(65536-46080)/256;    // 由于晶振为11.059 2MHz，故所记次数应为46 080次，计时器每隔50 000μs发起1次中断。
    TL0=(65536-46080)%256;    //46 080为50 000×11.059 2/12所得
    while(1)
    {
        if(k4 == 0)// 判断是否有按下按键的信号
        {
            delay(10);// 延时10ms，消抖
            if(k4 == 0)// 再次判断按键是否被按下
            {
                while(k4 == 0);// 直到判断按键松开
                {
                    if(stop)
                        start_timer();      // 打开定时器
                    else
                        stop_timer();       // 关闭定时器
                    stop = ~ stop;
                    beep = 1;
                    P1 = 0xff;
                }
            }
        }
        else if(k5 == 0)// 判断是否有按下按键的信号
        {
            delay(10);// 延时10ms，消抖
            if(k5 == 0)// 再次判断按键是否被按下
            {
                while(k5 == 0);// 直到判断按键松开
                {
                    stop_timer();// 关闭定时器
                    stop = 1;
                    beep = 1;
                    P1 = 0xff;
                    count = 0;
                    time_count = 0;
                    sec = 0;        // 计时清零
                    min = 0;
                }
            }
        }

        if(stop)
        {
            count++;
            if(count == 100)
            {
                P1 = ~ P1;
```

```
                count = 0;
            }
        }
        else if (time_count==20)         // 计数20次0.05s为1s
        {
            count = 0;
            time_count=0;
            sec++;
            if (sec==60)
            {
                sec=0;                    // 若到了60s, 则归零
                min++;
            }
            if (min==60)
            {
                min=0;                    // 若到了60min, 则归零
            }
            beep = 0;
            delay(10);
            beep = 1;
        }
        num0=sec%10;                     // 取出当前描述的个位与十位
        num1=sec/10;
        num2=min%10;                     // 取出当前描述的个位与十位
        num3=min/10;
        display(num3,num2,num1,num0);    // 显示
    }
}

void timer0()interrupt 1
{
    TH0= (65536-46080)/256;
    TL0= (65536-46080)%256;
    time_count++;
}
```

图 12-2 和图 12-3 所示是程序烧录之后电路板的状态，未按键盘启动秒表，IED 成闪烁状态。

图 12-2 串口通信实验实物图 1

图 12-3 串口通信实验实物图 2

图12-4、图12-5、图12-6所示是按了K4键，秒表启动实物图，每增加1s，蜂鸣器都会"滴"一声。

图12-4　串口通信实验结果实物图1

图12-5　串口通信实验结果实物图2

图12-6　串口通信实验结果实物图3

此项目附带视频及代码资料。

Chapter 13

项目13
LCD显示器静态显示字符

项目目标

- 通过学会如何进行LCD显示器（液晶屏）静态显示字符，为扩展项目打下基础。

建议学时

- 4学时。

知识要点

- LCD1620。
- LCD显示器字符的编写。
- 数据指针的设置。

技能掌握

- 掌握LCD1620的初始化及操作程序的编写。

13.1 项目分析

在单片机中,最常用的输出模块就是 LCD 显示器,本项目就选取了其典型程序的编制方法进行学习。

13.2 技术准备

13.2.1 LCD 1602 介绍

LCD1602 也叫 1602 字符型液晶屏,它是一种专门用来显示字母、数字、符号的点阵型液晶模块,由若干个 5×7 或者 5×11 的点阵字符位组成,每个点阵字符位都可以显示 1 个字符,每位之间有 1 个点距的间隔,每行之间也有间隔,即具有字符间距和行间距,也正是因为如此,LCD1602 不能很好地显示图片,实物图如图 13-1 和图 13-2 所示。

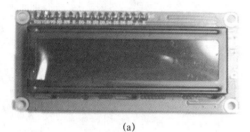
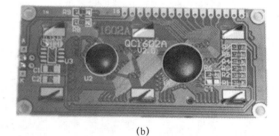

(a)　　　　　　　　　　　　　　　　(b)

图 13-1　LCD1602 实物图 1

图 13-2　LCD1602 液晶屏实物图 2

LCD1620 模块的原理图可以在本书最后附图 1 查看,LCD1602 的引脚见表 13-1。

表 13-1 LCD1602 引脚

编号	符号	引脚说明	编号	符号	引脚说明
1	VSS	电源地	9	D2	Data I/O
2	VDD	电源正极	10	D3	Data I/O
3	VL	液晶显示偏压信号	11	D4	Data I/O
4	RS	数据/命令选择端（H/L）	12	D5	Data I/O
5	R/W	读/写选择端（H/L）	13	D6	Data I/O
6	E	使能信号	14	D7	Data I/O
7	D0	Data I/O	15	BLA	背光源正极
8	D1	Data I/O	16	BLK	背光源负极

13.2.2 LCD1602 的驱动操作

1. LCD1602 的初始化程序

```
/*****************************************************************
* 函数名：LcdInit()
* 函数功能：初始化 LCD 显示器
* 输入 ：无
* 输出 ：无
******************************************************************/
void LCD_Init(void)
{
    LCD_Write_Com(0x38);    /*显示模式设置*/
    DelayMs(5);
    LCD_Write_Com(0x38);
    DelayMs(5);
    LCD_Write_Com(0x38);
    DelayMs(5);
    LCD_Write_Com(0x38);
    LCD_Write_Com(0x08);    /*显示关闭*/
    LCD_Write_Com(0x01);    /*显示清屏*/
    LCD_Write_Com(0x06);    /*显示光标移动设置*/
    DelayMs(5);
    LCD_Write_Com(0x0C);    /*显示开及光标设置*/
}
```

2. 写入命令程序

RS=L，RW=L，D0 ~ D7 为指令码，E 为高脉冲。

```
/*****************************************************************
* 函数名：LcdWriteCom
* 函数功能：向 LCD 写入 1 字节的命令
* 输入：com
* 输出：无
******************************************************************/
void LcdWriteCom(unsigned char com)   // 写入命令
{
    RS=0;
    RW=0;
    GPIO_LCD=com;
```

```
    Delay1ms(10);
    LCDE=1;
    Delay1ms(10);
    LCDE=0;
}
```

3. 写入数据程序

```
/*****************************************************************
* 函数名：LcdWriteData
* 函数功能：向 LCD 写入 1 字节的数据
* 输入：dat
* 输出：无
*****************************************************************/
void LcdWriteData (unsigned char dat)                // 写入数据
{
    RS=1;
    RW=0;
    GPIO_LCD=dat;
    Delay1ms(10);
    LCDE=1;
    Delay1ms(10);
    LCDE=0;
}
```

4. 写操作时序

写操作时序如表 13-2 所示。

表 13-2 写操作时序

时序参数	符号	极限值		
		最小值	典型值	最大值
E 信号周期 /ns	t_C	400	—	—
E 脉冲宽度 /ns	t_{pw}	150	—	—
E 上升沿/下降沿时间 /ns	t_R, t_F	—	—	25
地址建立时间 /ns	t_{SP1}	30	—	—
地址保持时间 /ns	t_{HD1}	10	—	—
数据建立时间（读操作）/ns	t_D	—	—	100
数据保持时间（读操作）/ns	t_{HD2}	20	—	—
数据建立时间（写操作）/ns	t_{SP2}	40	—	—
数据保持时间（写操作）/ns	t_{HD2}	10	—	—

13.2.3 LCD 1602 的指令码

（1）LCD1602 初始化见表 13-3。

表 13-3 初始化

指令码								功能
0	0	1	1	1	0	0	0	设置 16×2 显示、5×7 点阵、8 位数据接口

（2）LCD1602 开关显示及光标设置见表 13-4。

表 13-4　开、关显示及光标设置

指　令　码								功　　能
0	0	0	0	1	D	C	B	D=1 开显示；D=0 关显示。 C=1 显示光标；C=0 不显示光标。 B=1 光标闪烁；B=0 光标不显示。
0	0	0	0	0	1	N	S	N=1 当读或写 1 个字符后，地址指针加 1，且光标加 1。 N=0 当读或写 1 个字符后，地址指针减 1，且光标减 1。 S=1 当写 1 个字符后，整屏显示左移（N=1）或右移。 （N=0），以得到光标不移动而屏幕移动的效果。 S=0 当写 1 个字符后，整屏显示不移动。

（3）LCD1602 数据指针设置见表 13-5。

表 13-5　数据指针设置

指　令　码	功　　能
80H+ 地址码（0～27H），40H～67H）	设置数据地址指针

（4）LCD1602 其他设置见表 13-6。

表 13-6　其他设置

指　令　码	功　　能
01H	显示清屏包括数据指针清零和所有显示清零

13.2.4　RAM 地址映射图

RAM 地址映射如图 13-3 所示。

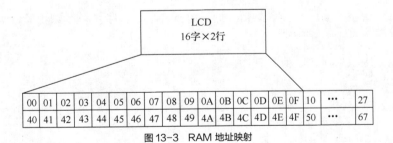

图 13-3　RAM 地址映射

13.3　项目实施

LCD1602 静态显示字符实验程序如下。

```
/**********************BST-V51 实验开发板例程 ***********************
 *  平台：BST-M51 + Keil U4 + STC89C51
 *  名称：LCD1602 模块实验
```

```
 *    晶振：11.059 2MHz
**********************************************************/
/*--------------------------------------------------
名称：LCD1602
内容：通过标准程序静态显示字符
引脚定义：1 VSS；2 VDD；3 V0；4 RS；5 R/W；6 E 7～14 DB0～DB7；15 BLA；16 BLK
------------------------------------------------*/
#include<reg52.h>  // 包含头文件，一般情况不需要改动，头文件包含特殊功能寄存器的定义
#include<intrins.h>

sbit RS = P1^0;    // 定义端口
sbit RW = P1^1;
sbit EN = P2^5;

#define RS_CLR RS=0
#define RS_SET RS=1

#define RW_CLR RW=0
#define RW_SET RW=1

#define EN_CLR EN=0
#define EN_SET EN=1

#define DataPort P0

/*------------------------------------------------
    μs 延时函数，含有输入参数 unsigned char t，无返回值 unsigned char 是定义无符号字符变量，其值的范围是
0～255 这里使用晶振 12MHz，精确延时请使用汇编，大致延时
长度为 T= t×2+5μs
------------------------------------------------*/
void DelayUs2x(unsigned char t)
{
    while(--t);
}
/*------------------------------------------------
    ms 延时函数，含有输入参数 unsigned char t，无返回值 unsigned char 是定义无符号字符变量，其值的范围是
0～255 这里使用晶振 12MHz，精确延时请使用汇编
------------------------------------------------*/
void DelayMs(unsigned char t)
{

    while(t--)
    {
        // 大致延时 1ms
        DelayUs2x(245);
        DelayUs2x(245);
    }
}
/*------------------------------------------------
判忙函数
------------------------------------------------*/
bit LCD_Check_Busy(void)
{
```

```
    RS_CLR;
    RW_SET;
    EN_CLR;
    _nop_();
    EN_SET;
    return (bit)(DataPort & 0x80);//0x80;0x00
    }
```
/*--
写入命令函数
--*/
```
    void LCD_Write_Com (unsigned char com)
    {
    while (LCD_Check_Busy()); //忙则等待
    RS_CLR;
    RW_CLR;
    EN_SET;
    DataPort= com;
    _nop_();
    EN_CLR;
    }
```
/*--
写入数据函数
--*/
```
    void LCD_Write_Data (unsigned char Data)
    {
    while (LCD_Check_Busy()); //忙则等待
    RS_SET;
    RW_CLR;
    EN_SET;
    DataPort= Data;
    _nop_();
    EN_CLR;
    }
```

/*--
清屏函数
--*/
```
    void LCD_Clear (void)
    {
    LCD_Write_Com(0x01);
    DelayMs(5);
    }
```
/*--
写入字符串函数
--*/
```
    void LCD_Write_String (unsigned char x,unsigned char y,unsigned char *s)
    {
    if (y == 0)
        {
        LCD_Write_Com(0x80 + x);        //表示第1行
        }
    else
        {
        LCD_Write_Com(0xC0 + x);        //表示第2行
```

```c
        }
    while (*s)
        {
        LCD_Write_Data ( *s);
        s ++;
        }
    }
/*------------------------------------------------
写入字符函数
------------------------------------------------*/
    void LCD_Write_Char (unsigned char x,unsigned char y,unsigned char Data)
    {
    if (y == 0)   // 设置坐标
        {
        LCD_Write_Com (0x80 + x);
        }
    else
        {
        LCD_Write_Com (0xC0 + x);
        }
    LCD_Write_Data ( Data);
    }
/*------------------------------------------------
初始化函数
------------------------------------------------*/
    void LCD_Init (void)
    {
    LCD_Write_Com (0x38);       /*显示模式设置*/
    DelayMs (5);
    LCD_Write_Com (0x38);
    DelayMs (5);
    LCD_Write_Com (0x38);
    DelayMs (5);
    LCD_Write_Com (0x38);
    LCD_Write_Com (0x08);       /*显示关闭*/
    LCD_Write_Com (0x01);       /*显示清屏*/
    LCD_Write_Com (0x06);       /*显示光标移动设置*/
    DelayMs (5);
    LCD_Write_Com (0x0C);       /*显示开及光标设置*/
    }

/*------------------------------------------------
主函数
------------------------------------------------*/
void main (void)
{
LCD_Init();
LCD_Clear();// 清屏
while (1)
    {
    LCD_Write_Char (7,0, 'o');
    LCD_Write_Char (8,0, 'k');
    LCD_Write_String (1,1, "hello world");
while (1);
```

    }
}

LCD1602 静态显示字符的实际操作结果实物图如图 13-4 所示。

图 13-4　LCD1602 静态显示字符实验结果实物图

注意，LCD1602 接口直接插在开发板上，不需要杜邦线连接。

此项目附操作视频及代码资料。

Chapter 14

项目14
红外遥控

项目目标

- 通过学习基本的红外发送与接收，为后面综合实验项目打下基础。

建议学时

- 8学时。

知识要点

- 红外线。
- 遥控器。

技能掌握

- 学会红外线发射模块的程序编写，及其红外解码的程序编写，并熟知整个发射、接收流程。

14.1 项目分析

通用红外遥控系统由发射和接收两大部分组成。应用编/解码专用集成电路芯片来进行控制操作。发射部分包括键盘矩阵、编码调制、LED 红外发送器；接收部分包括光、电转换放大器、解调、解码电路。

14.2 技术准备

14.2.1 红外线

1. 红外线及其应用简介

（1）红外线简介。

在光谱中波长在 760nm ～ 400μm 的电磁波称为红外线，它是一种不可见光。目前几乎所有的视频和音频设备都可以通过红外遥控的方式进行遥控，比如电视机、空调、影碟机等。这种技术应用广泛，相应的应用器件都十分廉价，因此红外遥控是设备控制的理想方式。红外线原理图可在项目 2 图 2-4、图 2-5 查看。

（2）红外线有以下几个主要的应用领域

① 红外线辐射加热。比如理疗机就使用了远红外线的热效应，治疗某些疾病。

② 红外线测温。比如夜视仪就是通过探测人体热量，进行红外线成像的。又比如著名的美国响尾蛇导弹就是一种跟踪飞机尾部热量的热寻导弹。

③ 红外线通信。比如测距仪就是以红外线作为载波的一种测量距离的精密仪器。再比如红外遥控器也是以红外线作为载波的一种无线通信设备。

（3）红外遥控的优点及应用场合。

红外线遥控是目前使用最广泛的一种通信和遥控手段。红外线遥控装置的优点是体积小、功耗低、功能强、成本低。在家用电器如彩电、录像机、音响设备、空调机以及玩具等产品中应用非常广泛。工业设备中，在高压、辐射、有毒气体、粉尘等环境下，采用红外线遥控，不仅完全可靠而且能有效地隔离电气干扰。

（4）红外二极管包括以下 2 种。

① 红外发光二极管通常使用砷化镓（GaAs）、砷铝化镓（GaAlAs）等材料，采用全透明或浅蓝色、黑色的树脂封装。产生的光波波长为 940nm 左右，为红外光，如图 14-1 所示。

② 红外接收头如图 14-2 所示，其内部含有高频的滤波电路，专门用来滤除红外线合成信号的载波信号（38kHz），并送出接收到的信号。当红外线合成信号进入红外接收模块，在其输出端便可以得到原先发射器发出的数字编码，只要经过单片机解码程序进行解码，便可以得知按下了哪一个按键，而做出相应的控制处理，完成红外遥控的动作。

图 14-1 红外发光二极管

图 14-2 红外接收头

红外接收头的主要参数如下。

工作电压：4.8 ~ 5.3V。

工作电流：1.7 ~ 2.7mA。

接收频率：38kHz。

峰值波长：980nm。

静态输出：高电平。

输出低电平：≤ 0.4V。

输出高电平：接近工作电压。

2. 红外遥控系统结构

通用红外遥控系统由发射和接收两大部分组成，应用编 / 解码专用集成电路芯片进行控制操作。发射部分包括键盘、编码调制、LED 红外发送器；接收部分包括光 / 电转换放大器、解调、解码电路。

红外遥控系统框图如图 14-3 所示。

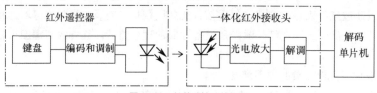

图 14-3　红外遥控系统框图

（1）发射系统。

① 遥控发射器及其编码电路结构示意图如图 14-4 所示。

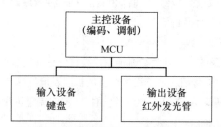

图 14-4　遥控发射器及其编码电路结构示意图

② 硬件电路的设计。

硬件调制发射电路如图 14-5 所示。

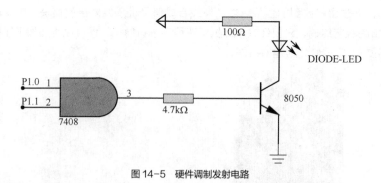

图 14-5　硬件调制发射电路

软件调制发射电路如图 14-6 所示。

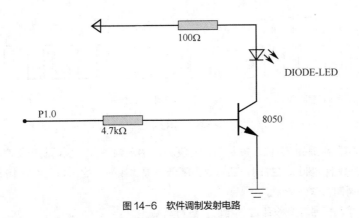

图 14-6　软件调制发射电路

③ 各种编码的作用

引导码：单片机只有检测到了引导码才确认接收后面的数据，保证数据接收的正确性。

客户码：为了区分各红外遥控设备，使之不会互相干扰。

操作码：用户实际需要的编码，按下不同的键，产生不同的操作码，待接收端接收到后，根据其进行不同的操作。

操作反码：为操作码的反码，目的是接收端接收到所有数据之后，将其取反，与操作码比较，不相等，则表示在传输过程中编码发生了变化，此次接收的数据视为无效，可提高接收数据的准确性。

a. 引导码的波形定义。

（a）一般的红外发射芯片，比如日本 NEC 的 uPD6121G 定义的引导码为，9ms 高电平 +4.5ms 低电平（4.5ms 低电平也叫结果码），如图 14-7 所示。

（b）引导码也可以自定义（为了接收准确，引导码高电平状态时间不能过短）。

b. 客户码、操作码的波形定义。

（a）客户码和操作码都为 8 位二进制编码。

（b）日本 NEC 的 uPD6121G 定义的 "0" 和 "1" 如下。

"0" 0.56ms 高电平 + 0.565ms 低电平。

"1" 0.56ms 高电平 + 1.685ms 低电平。

如图 14-8 所示。同样，这样的数码 "0" 和 "1" 的占空比也可以自定义。

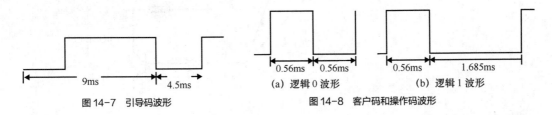

图 14-7　引导码波形　　　　图 14-8　客户码和操作码波形

④ 调制。

现在假如要发送一个数据 ECH，其客户码 1 为 AAH，客户码 2 为 66H，则发送的二进制数为 01010101　01100110　00010011　11101100

再加上引导码，要发送的波形如图 14-9 所示。这样传输信号的过程就称为调制。

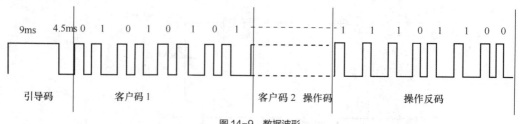

图 14-9　数据波形

上述"0"和"1"组成的 32 位二进制码，经 38kHz 的载频进行二次调制以提高发射效率（因红外接收头能接收的红外线为 38kHz 左右），还可达到降低电源功耗的目的，如图 14-10 所示。

调制分为硬件调制与软件调制。

硬件调制：将编码信号与载波通过"与门"进行调制。

软件调制：直接用软件产生调制后的信号。

如图 14-11 和图 14-12 所示。

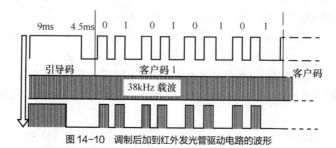

图 14-10　调制后加到红外发光管驱动电路的波形

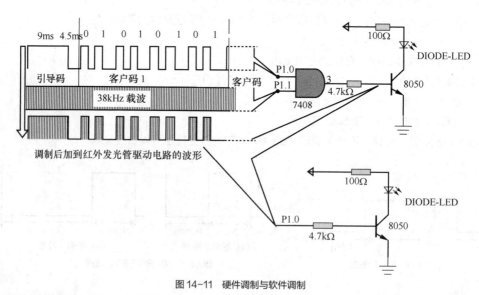

图 14-11　硬件调制与软件调制

调制的程序框图如图 14-12 所示。

（2）接收系统。

① 红外接收及解码电路结构示意图如图 14-13 所示。

② 红外线接收电路如图 14-14 所示。

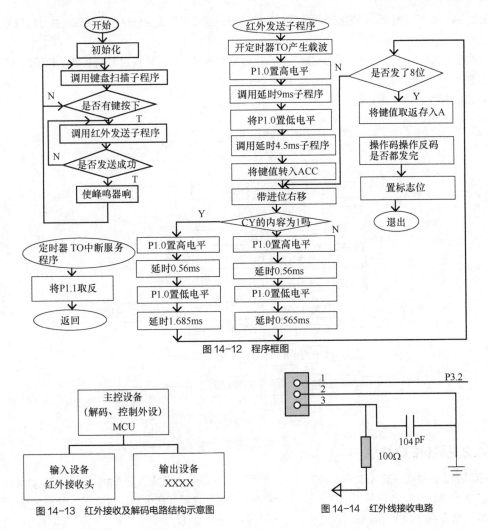

图 14-12 程序框图

图 14-13 红外接收及解码电路结构示意图

图 14-14 红外线接收电路

③ 红外接收头的特性。

当接收到 38kHz 的红外线时其输出低电平;静态时其输出为高电平,如图 14-15 所示。

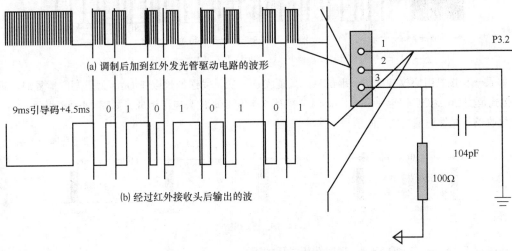

图 14-15 红外接收头静态时输出高电平

由图 14-15 可以看出，经过红外发光管发出的信号，经红外接收头已进行了解调，并且将信号进行了取反。

同时还可以看出，"0"和"1"只是低电平时间长度不同，因此可通过时间长度来判断是"0"还是"1"，如图 14-16（a）所示。

④ 红外接收程序框图如下。

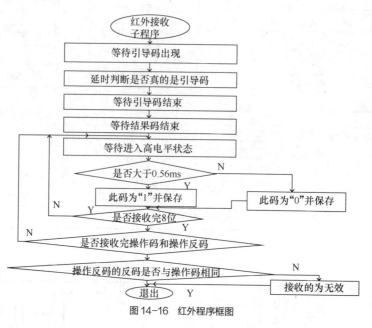

图 14-16　红外程序框图

14.2.2　NEC 协议

NEC 协议规定低位首先发送。如图 14-17 所示，一串信息首先发送 9ms AGC（自动增益控制）高脉冲，然后发送 4.5ms 起始低电平，接着发送 4 字节的地址码和命令码。这 4 字节分别为地址码、地址码反码、命令码、命令码反码。

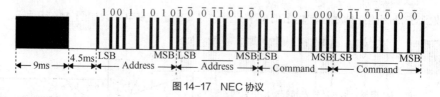

图 14-17　NEC 协议

如果一直按发射协议按键，一串信息也只能发送一次，发送的是以 110ms 为周期的重复码。重复码的格式是由 9msAGC 高电平和 4.5ms 低电平及 1 个 560us 的高电平组成，如图 14-18 所示。

接收信号是发送信号的反向信号。

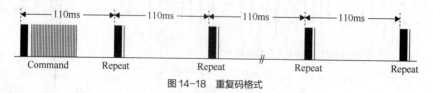

图 14-18　重复码格式

程序解析图如图 14-19 所示，程序解析过程如下。

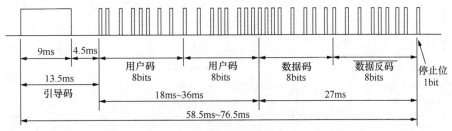

图 14-19　程序解析图

产生下降沿，进入外部中断 0 的中断函数，延时之后，检测 I/O 端口是否还是高电平，是，就等待 9ms 的高电平结束。

待 9ms 高电平结束，继续等待 4.5ms 低电平结束。然后开始接收传送的 4 组数据。

先等待 0.560ms 高电平结束。接着检测高电平的持续时间，如果超过 1.12ms，那么是低电平（低电平的的持续时间为 1.69ms，高电平的持续时间为 0.565ms）。最后将接收到的数据和数据的反码进行比较，检测其是否相同。

14.2.3　红外遥控器键码值

红外遥控器键码值如图 14-20 所示。

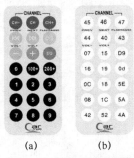

图 14-20　红外遥控器键码值

14.3　项目实施

14.3.1　红外线发送

红外线发送模块实验程序如下。

```
/********************BST-V51 实验开发板例程 **********************
*   平台：BST-M51 + Keil U4 + STC89C51
*   名称：红外发送模块实验
*   晶振：11.059 2MHz
***************************************************/
#include <reg52.h>

#define uchar unsigned char
#define uint unsigned int
```

```c
sbit k1 = P3^4;// 定义 4 个独立按键
sbit k2 = P3^5;
sbit k3 = P3^6;
sbit k4 = P3^7;
sbit out=P1^5;// 发送 I/O 端口
const uchar TabHL1[12]={0x30,0x18,0x7a,0x10,0x38,0x5a,0x42,0x4a,
                        0x52,0x00,0xff,0xa6};// 数据码码表 1～9 及 2 字节用户码
/*----------------------------------------------
延时函数
----------------------------------------------*/
void delay(uint xms)
{
  uint i,j;
  for(i=xms;i>0;i--)        //i=xms 即延时 xms
     for(j=112;j>0;j--);
}

void delay560us(void) //0.56ms 延迟函数
{
  uint j;
  for(j=63;j>0;j--);
}

void delay4500us(void)    //4.5ms 延迟函数
{
  uint j;
  for(j=516;j>0;j--);
}

void khz_2(uint num)            //38kHz 脉冲占空比 1:2
{
   for(;num>0;num--)
   {
//   _nop_();
     out=~out;
   }
}

/*----------------------------------------------
发送逻辑 0
----------------------------------------------*/
void send0(void)
{
   khz_2(42);
   //khz_3(21);
   out=1;
   delay560us();
}
/*----------------------------------------------
发送逻辑 1
----------------------------------------------*/
void send1(void)
{
   khz_2(42);
```

```c
    //khz_3(21);
    out=1;
    delay560us();
    delay560us();
    delay560us();
}
/*------------------------------------------------
发送1字节数据
------------------------------------------------*/
void Send8Bit (uchar dat)
{
   if(dat&0x80){    send1();}
   else{            send0();}
   if(dat&0x40){    send1();}
   else{            send0();}
   if(dat&0x20){    send1();}
   else{            send0();}
   if(dat&0x10){    send1();}
   else{            send0();}
   if(dat&0x08){    send1();}
   else{            send0();}
   if(dat&0x04){    send1();}
   else{            send0();}
   if(dat&0x02){    send1();}
   else{            send0();}
   if(dat&0x01){    send1();}
   else{            send0();}

}


/*------------------------------------------------
发送用户码
------------------------------------------------*/
void usercode()    // 发送用户码 00FF
{
    Send8Bit(TabHL1[9]);
    Send8Bit(TabHL1[10]);
}
/*------------------------------------------------
发送引导码
------------------------------------------------*/
void leadcode(void) // 发送引导码
{
    khz_2(690);
    //khz_3(345);
    out=1;
    delay4500us();

}
/*------------------------------------------------
发送红外值
------------------------------------------------*/
void send_1()//00110000              发送"1"的信号
   {
       leadcode();
       usercode();
```

```c
        Send8Bit (TabHL1[0]);
        Send8Bit (~TabHL1[0]);
    }

    void send_2()//00011000    发送"2"的信号
    {
        leadcode();
        usercode();
        Send8Bit (TabHL1[1]);
        Send8Bit (~TabHL1[1]);
    }

    void send_3()//01111010    发送"3"的信号
    {
        leadcode();
        usercode();
        Send8Bit (TabHL1[2]);
        Send8Bit (~TabHL1[2]);
    }

    void send_4()//00010000    发送"4"的信号
    {
        leadcode();
        usercode();
        Send8Bit (TabHL1[3]);
        Send8Bit (~TabHL1[3]);
    }

/*-------------------------------------------------
按键扫描函数
-------------------------------------------------*/
void keyscan()
{
    while (1)
    {
        if (k1 == 0)// 判断是否有按下按键的信号
        {
            delay(10);// 延时10ms 消抖
            if (k1 == 0)// 再次判断按键是否被按下
            {
                while (k1 == 0);// 直到判断按键松开
                send_1();// 松开后执行程序发送"1"的信号
            }
        }
        if (k2 == 0)// 判断是否有按下按键的信号
        {
            delay(10);// 延时10ms 消抖
            if (k2 == 0)// 再次判断按键是否被按下
            {
                while (k2 == 0);// 直到判断按键松开
                send_2();// 松开后执行程序发送"2"的信号
            }
        }
        if (k3 == 0)// 判断是否有按下按键的信号
        {
            delay(10);// 延时10ms 消抖
```

```c
            if(k3 == 0)// 再次判断按键是否被按下
            {
                while(k3 == 0);// 直到判断按键松开
                send_3();// 松开后执行程序发送"3"的信号
            }
        }
        if(k4 == 0)// 判断是否有按下按键的信号
        {
            delay(10);// 延时10ms 消抖
            if(k4 == 0)// 再次判断按键是否被按下
            {
                while(k4 == 0);// 直到判断按键松开
                send_4();// 松开后执行程序发送"4"的信号
            }
        }
    }
}

/*----------------------------------------------
主函数
----------------------------------------------*/
void main(void)
{
    while(1)
    {
        keyscan();
    }
}
```

详细操作过程可见附带视频。

14.3.2 红外线接收

红外线接收模块实验程序如下。

```c
/*********************BST-V51 实验开发板例程 **********************
* 平台：BST-M51 + Keil U4 + STC89C51
* 名称：红外接收模块实验
* 晶振：11.059 2MHz
*************************************************************/
/*----------------------------------------------
名称：遥控器红外解码LED显示器显示
内容：按配套遥控器上1～9键，会在LED显示器上对应显示
----------------------------------------------*/
#include<reg52.h>        // 包含头文件，一般情况不需要改动，头文件包含特殊功能寄存器的定义

sbit IR=P3^2;    // 红外接口标志

#define DataPort P0 // 定义数据端口程序中遇到DataPort端口则用P0引脚替换
sbit wei1 = P2^4;// 定义第1位LED显示器
sbit wei2 = P2^5;// 定义第2位LED显示器
sbit wei3 = P2^6;// 定义第3位LED显示器
sbit wei4 = P2^7;// 定义第4位LED显示器
/*----------------------------------------------
全局变量声明
----------------------------------------------*/

unsigned char code DuanMa[10]={0x3f,0x06,0x5b,0x4f,0x66,0x6d,0x7d,
                                0x07,0x7f,0x6f};// 显示段码值0～9
```

```c
unsigned char  irtime;//红外用全局变量

bit irpro_ok,irok;
unsigned char IRcord[4];
unsigned char irdata[33];

/*------------------------------------------------
函数声明
------------------------------------------------*/

void Ir_work(void);
void Ircordpro(void);

/*------------------------------------------------
定时器T0中断处理
------------------------------------------------*/

void tim0_isr (void) interrupt 1 using 1
{
  irtime++;   //用于计数2个下降沿之间的时间
}


/*------------------------------------------------
外部中断0中断处理
------------------------------------------------*/
void EX0_ISR (void) interrupt 0 //外部中断0服务函数
{
    static unsigned char  i;               //接收红外信号处理
    static bit startflag;                  //是否开始处理标志位

    if(startflag)
    {
        if(irtime<63&&irtime>=33)//引导码 TC9012的头码,9ms+4.5ms
        i=0;
        irdata[i]=irtime;//存储每个电平的持续时间,用于以后判断是0,还是1
        irtime=0;
        i++;
        if(i==33)
        {
         irok=1;
         i=0;
        }
    }
    else
    {
        irtime=0;
        startflag=1;
    }

}

/*------------------------------------------------
定时器T0初始化
------------------------------------------------*/
void TIM0init (void)//定时器T0初始化
{
```

```c
    TMOD=0x02;// 定时器 T0 工作方式 2，TH0 是重装值，TL0 是初值
    TH0=0x00;   // 重载值
    TL0=0x00;   // 初始化值
    ET0=1;      // 开中断
    TR0=1;
}
/*------------------------------------------------
外部中断 0 初始化
------------------------------------------------*/
void EX0init (void)
{
  IT0 = 1;              // 指定外部中断 0 下降沿触发，INT0 端口（P3.2 引脚）
  EX0 = 1;              // 使能外部中断
  EA = 1;               // 开总中断
}
/*------------------------------------------------
键值处理
------------------------------------------------*/

void Ir_work (void) // 红外键值散转程序
{
        switch (IRcord[2])// 判断第 3 个数码值
           {
            case 0x0c:DataPort=DuanMa[1];break;//1 显示相应的按键值
            case 0x18:DataPort=DuanMa[2];break;//2
            case 0x5e:DataPort=DuanMa[3];break;//3
            case 0x08:DataPort=DuanMa[4];break;//4
            case 0x1c:DataPort=DuanMa[5];break;//5
            case 0x5a:DataPort=DuanMa[6];break;//6
            case 0x42:DataPort=DuanMa[7];break;//7
            case 0x52:DataPort=DuanMa[8];break;//8
            case 0x4a:DataPort=DuanMa[9];break;//9
            default:break;
           }

       irpro_ok=0;// 处理完成标志

 }
/*------------------------------------------------
红外码值处理
------------------------------------------------*/
void Ircordpro (void) // 红外码值处理函数
{
  unsigned char i, j, k;
  unsigned char cord,value;

  k=1;
  for (i=0;i<4;i++)        // 处理 4 字节
     {
      for (j=1;j<=8;j++) // 处理 1 字节 8 位
          {
           cord=irdata[k];
           if (cord>7) // 大于某值为 1，这个和晶振有绝对关系，这里使用 12MHz 计算，此值可以有一定误差
              value|=0x80;
           if (j<8)
```

```c
            {
             value>>=1;
            }
            k++;
         }
      IRcord[i]=value;
      value=0;
   }
    irpro_ok=1;// 处理完毕标志位置1
}


/*---------------------------------------------
主函数
---------------------------------------------*/
void main (void)
{
  EX0init(); // 初始化外部中断
  TIM0init();// 初始化定时器

  wei1 = 1;  // 第1位LED显示器选通
  wei2 = 1;  // 第2位LED显示器选通
  wei3 = 1;  // 第3位LED显示器选通
  wei4 = 1;  // 第4位LED显示器选通

while (1) // 主循环
  {
    if (irok)                    // 如果接收好了,进行红外处理
    {
     Ircordpro();
     irok=0;
    }

    if (irpro_ok)                // 如果处理好后,进行工作处理,如按对应的按键后,显示对应的数字等
    {
     Ir_work();
    }
   }
}
```

此项目附操作视频及代码资料。

注意,附带资料对应项目有遥控器的代码设置方法文件。

Chapter 15

项目15
简易计算器（LED显示器显示）

项目目标
- 了解矩阵键盘检测原理及如何获得键盘扫描。
- 掌握矩阵键盘的检测和LED显示器显示混合编程。

建议学时
- 4学时。

知识要点
- 4×4 矩阵的工作原理。
- LED显示器显示混合编程。

技能掌握
- 学会利用51芯片的I/O引脚，完成系统硬件电路设计。

15.1 项目分析

矩阵键盘是单片机外部设备中所使用的、排布类似于矩阵的键盘组。设计制作一个检测 4×4 矩阵键盘的按键编码实验,把实际按键键值的 8 位编码先转换成 0000 ~ 1111 的编码,再译成 LED 显示器能识别的 8 位编码。

15.2 技术准备

矩阵按键部分由 16 个轻触按键按照 4 行 4 列排列,连接到 JP50 端口。将行线所接的单片机的 I/O 端口作为输出端,而列线所接的 I/O 端口则作为输入端。这样,当按键未按下时,所有的输出端都是高电平,代表无键按下,行线输出低电平。一旦有键按下,则输入线就会被拉低,这样,通过读入输入线的状态就可得知是否有键按下了。矩阵键盘实物及其内部接线如图 15-1 所示。以下是 2 种键盘扫描方式。

(1)逐行扫描:可以通过高 4 位轮流输出低电平,来对矩阵键盘进行逐行扫描,当低 4 位接收到的数据不全为高电平时,说明有按键按下,然后通过接收到的数据是哪一位为 0,来判断是哪一个按键被按下。

(2)行列扫描:可以通过高 4 位全部输出低电平,低四位输出高电平。当接收到的数据,低四位不全为高电平时,说明有按键按下,然后通过接收的数据值,判断是哪一列有按键按下,然后再反过来,高四位输出高电平,低四位输出低电平,然后根据接收到的高四位的值判断是那一行有按键按下,这样就能够确定是哪一个按键按下了。

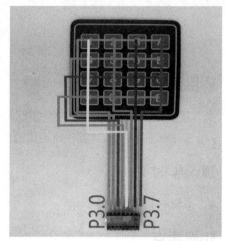

图 15-1 矩阵键盘实物及其内部接线图

15.3 项目实施

矩阵键盘扫描 LED 显示器显示实验程序如下。实验实物连接图如图 15-2 所示。

```
/*******************BST-M51 实验开发板例程 *************
*   平台:BST-M51 + Keil U4 + STC89C51
*   名称:矩阵键盘扫描 LED 显示器显示
*   晶振:11.059 2MHz
*   实验效果:按 4×4 矩阵按键,LED 显示器 4 位依次显示1,2,3,A;4,5,6,B;7,8,9,C;E,0,F,D
*****************************************************/
#include<reg52.h>
#define uchar   unsigned char
#define uint    unsigned int
uchar code table[17]={0x3f , 0x06 , 0x5b ,0x4f , 0x66 , 0x6d ,0x7d ,
                      0x07, 0x7f , 0x6f ,0x77 , 0x7c , 0x39 ,
                      0x5e , 0x79 , 0x71 , 0x00};    //0~9与A~F及"不显示"字型码

void delay(uint xms)// 函数延迟 xms
{
```

```c
    uint i,j;
    for(i=xms;i>0;i--)
    for(j=112;j>0;j--);
}

void display(uchar num)    //LED 显示器显示函数
{
    P0=table[num];                // 段选
}

void keyscan4x4()
{
    uchar temp,key;

    /////////////// 第 1 行扫描 ///////////////////
    P3=0xfe;//1111 1110 让 P3.0 引脚端口输出低电平
    temp=P3;
    temp=temp&0xf0;//1111 0000 位"与"操作屏蔽后 4 位
    if(temp!=0xf0)
    {
        delay(10);
        temp=P3;
        temp=temp&0xf0;
        if(temp!=0xf0)
        {
            temp=P3;
            switch(temp)
            {
                case 0xee:                //1110 1110 "1"被按下
                    key=1;
                    break;
                case 0xde:                //1101 1110 "2"被按下
                    key=2;
                    break;
                case 0xbe:                //1011 1110 "3"被按下
                    key=3;
                    break;
                case 0x7e:                //0111 1110 "A"被按下
                    key=10;
                    break;
            }
            while(temp!=0xf0)
            {
                temp=P3;
                temp=temp&0xf0;
            }
            display(key);
        }
    }
```

其他 3 行代码类似第 1 行。区别在于接口的改变。主函数如下。

```c
void main()
{
    P2 = P2 | 0xf0;              // 位选锁存为 4 位同时显示
    while(1)
```

```
    {
        keyscan4x4();
    }
}
```

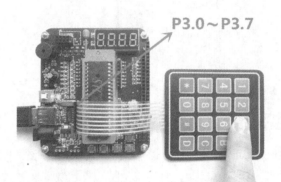

图 15-2　矩阵键盘扫描实验实物连接图

15.4　技术拓展

LED 显示器显示简易计算程序如下。

```
/*******************BST-M51 实验开发板例程 **********************
*   平台:BST-M51 + Keil U4 + STC89C51
*   名称:计算器
*   晶振:11.059 2MHz
*****************************************************************
键盘上按键说明:
0～9 数字输入;# 清零;A 加 ; B 减 ; C 除;D 等于
程序功能:本程序为简易计算器,可以算整数且正数类型的运算
*****************************************************************/
#include<reg51.h>

#define dula P0           // 段选信号的锁存器控制
#define wela P2           // 位选信号的锁存器控制,这里只用到 P2.4～P2.7 引脚

unsigned char temp,key=0,wei=0,i,j,k,keydown=0;
unsigned char jia=0,jian=0,cheng=0,chu=0,jia0=0,jian0=0,cheng0=0,chu0=0,dengyu=0,qingling=0;
unsigned char s0=16,s1=16,s2=16,s3=16;  // 参加运算的各个位
unsigned long qian = 0,hou = 0;// 定义参与运算的第 1 个数和第 2 个数。

unsigned char code table[]={0x3f,0x06,0x5b,0x4f,0x66,0x6d,0x7d,
                            0x07,0x7f,0x6f,0x77,0x7c,0x39,0x5e,0x79,0x71,0x00,0x40};
unsigned char code weitable[]={0x8f,0x4f,0x2f,0x1f};
                              //LED 显示器各位的码表
void delay(unsigned char i)
{
    for (j=i;j>0;j--)
    for (k=125;k>0;k--);
```

```
}
void display1(unsigned char wei,unsigned char shu)// 在任意一位显示任意的数字
{
    wei=wei-1;
    wela|=0xf0;// 给 P2.4～P2.7 引脚端口置 1
    P0=table[shu];
    wela=wela&weitable[wei];// 给 P2 需要显示的那一位置 1, 其他清 0
    delay(5);
}
void display(unsigned char a,unsigned char b,unsigned char c,unsigned char d)
{                                                //1 次显示 4 个数字，且每位显示数字可自定
display1(4,a);
    display1(3,b);
    display1(2,c);
    display1(1,d);
}

void keyscan4x4()
{
    unsigned char temp;

    //////////////// 第 1 行扫描 ////////////////////
    参考 15.3.1
    //////////////// 第 2 行扫描 ////////////////////
    参考 15.3.1
    //////////////// 第 3 行扫描 ////////////////////
    参考 15.3.1
    //////////////// 第 4 行扫描 ////////////////////
    参考 15.3.1
```

LED 显示器显示简易计算器实验操作结果实物图如图 15-3 所示。

图 15-3　简易计算器实验操作结果实物图

此项目附操作视频及代码资料。

Chapter 16

项目16 音乐喷泉

项目目标

- 通过制做音乐喷泉的软、硬件的实验,掌握用声音传感器检测环境声音强度,并对声音传感器进行调节。

建议学时

- 4学时。

知识要点

- 检测周围环境的声音强度。
- 接入声音传感器。

技能掌握

- 掌握声音传感器的调节方法。

16.1 项目分析

音乐喷泉主要是通过声音传感器实现的。声音传感器的作用相当于一个话筒（麦克风），用来接收声波，显示声音的振动图像。但不能对噪声的强度进行测量。

声音传感器内置一个对声音敏感的电容式驻极体话筒，声波使话筒内的驻极体薄膜振动，导致电容的变化，而产生与之对应变化的微小电压。这一电压随后被转化成 0~5V 的电压，经过 A/D 转换，被数据采集器接收，并传送给计算机。

16.2 技术准备

16.2.1 模块原理图

声音传感器模块原理图如图 16-1 所示。关于声音传感器模块需要注意以下几点。

（1）可以检测周围环境的声音强度。使用时要注意，此传感器只能识别声音的有无（根据震动原理），不能识别声音的大小，或者特定频率的声音。

（2）灵敏度可调（蓝色数字电位器调节）。

（3）工作电压 3.3~5V。

（4）输出形式为数字开关量输出（"0"和"1"）。

（5）设有固定螺栓孔，方便安装。

（6）小板 PCB（印制电路板）尺寸为 3.4cm×1.6cm。

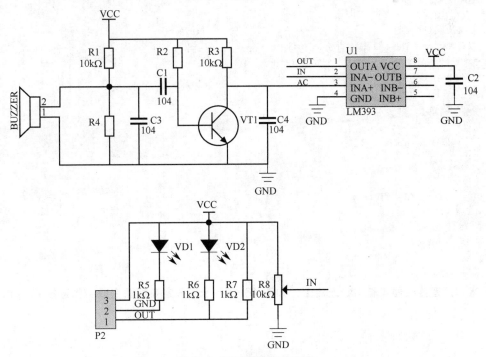

图 16-1　声音传感器模块原理图

16.2.2 模块接口说明

声音传感器模块接口说明如图 16-2 所示。
（1）VCC：外接 3.3～5V 电压（可以直接与 5V 单片机和 3.3V 单片机相连）。
（2）GND：外接 GND。
（3）DO（模块数字接口）：小板数字量输出接口（0 和 1）。

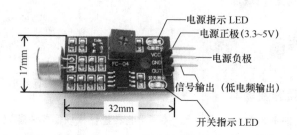

图 16-2 声音传感器模块接口说明

16.3 项目实施

声音传感器模块实验程序如下。

```c
/********************BST-M51 实验开发板例程 ********************
*   平台: BST-M51 + Keil U4 + STC89C51
*   名称: 声音模块检测
*   日期: 2015-6
*   晶振: 11.059 2MHz
*   实验效果: 有声音时，检测到，并使 LED1 亮
*************************************************/
#include<reg52.h>

sbit led1 = P1^0;//RED

sbit voice = P2^0;

void main()
{
  P1 = 0xff;// 熄灭所有 LED
  while(1)
  {
    led1 = voice;
  }
}
```

注意，声音传感器实验实物连接图，如图 16-3 所示。一定要按图完成实物连接后，才能进行操作。

图 16-3　声音传感器实验实物连接图

16.4 技术拓展

音乐喷泉实验程序如下。

```
/********************BST-M51 实验开发板例程 *********************
*  平台：BST-M51 + Keil U4 + STC89C51
*  名称：声音模块检测
*  晶振：11.059 2MHz
*  实验效果：有声音时，检测到，并使LED1亮
**********************************************/
#include<reg52.h>

sbit led1 = P1^0;//RED
sbit led2 = P1^1;//RED
sbit led3 = P1^2;//GREEN
sbit led4 = P1^3;//GREEN
sbit led5 = P1^4;//YELLOW
sbit led6 = P1^5;//YELLOW
sbit led7 = P1^6;//BLUE
sbit led8 = P1^7;//BLUE

sbit voice = P2^0;

unsigned long level = 0;

void timer0_init()
{
  TMOD = 0x01;  // 定时器T0选择工作方式1
    TH0 = 0xDC;      // 设置初始值10ms
    TL0 = 0x00;
    EA = 1;          // 打开总中断
    ET0 = 1;         // 打开定时器T0中断
    TR0 = 1;         // 启动定时器T0
}

void main()
```

```c
{
    P1 = 0xff;// 熄灭所有 LED
    timer0_init();
    while(1) // 根据不同等级判断 LED 亮
    {
      if(voice == 0)
      {
        led1 = voice;
      }
      if((led1 == 0) && (level >= 1))
      {
        led2 = 0;
      }
      if((led2 == 0) && (level >= 2))
      {
        led3 = 0;
      }
      if((led3 == 0) && (level >= 3))
      {
        led4 = 0;
      }
      if((led4 == 0) && (level >= 4))
      {
        led5 = 0;
      }
      if((led5 == 0) && (level >= 5))
      {
        led6 = 0;
      }
      if((led6 == 0) && (level >= 6))
      {
        led7 = 0;
      }
      if((led7 == 0) && (level >= 7))
      {
        led8 = 0;
      }
      if(level == 0)
      {
        P1 = 0xff;
      }
    }
}

void timer0()interrupt 1
{
    TH0 = 0xDC;        // 设置初始值 10ms
    TL0 = 0x00;
    if(voice == 0) // 每过 10ms 检测 1 次,如果还是有声音,则上升等级
        level++;
    else
        level = 0;
}
```

此项目附操作视频及代码资料。

Chapter 17

项目17 防盗报警器

项目目标

- 了解滚珠开关的工作原理,并掌握用滚珠开关制作的防盗报警器的软、硬件实现。

建议学时

- 4学时。

知识要点

- 认识滚珠开关。
- 用滚珠开关制作防盗报警器。

技能掌握

- 安装USB驱动;利用Keil C开发环境编辑、编译、调试C51程序的初步过程;掌握实用程序烧录方法以及相关工具。

17.1 项目分析

滚珠开关也叫钢珠开关、珠子开关，是震动开关的一种，是通过珠子滚动接触导针的原理，来控制电路的接通或者断开的。

17.2 项目准备

防盗报警器的主要部件就是滚珠开关。滚珠开关的工作原理和普通的电灯开关相似，开关触头触碰内部的金属板，电灯就亮，分开，电灯就灭。滚珠开关是利用开关中小珠的滚动，产生与金属端子的触碰或改变光线行进的路线，产生导通或不导通的效果。滚珠开关实物及原理图如图 17–1 所示，其运用层面极广，如胎压监控系统（TPMS）、脚踏车灯、数码相框旋转、银幕旋转、视讯镜头翻转、防盗系统等，举凡想侦测物体角度变化、倾倒、移动、震动、旋转的场合，滚珠开关皆适用。

滚珠开关，在电子玩具、医疗电子产品、车载音响、自动化控制系统、检测测量设备、汽车交通运输等有广泛的应用。作为滚珠最主要的应用领域，玩具行业、家用电器行业在我国得到了长足的发展。滚珠开关由于体积小重量轻在许多新兴领域得到广泛的应用，如数码相框的旋转屏幕、手机重力感应、防盗器材、智能化系统。

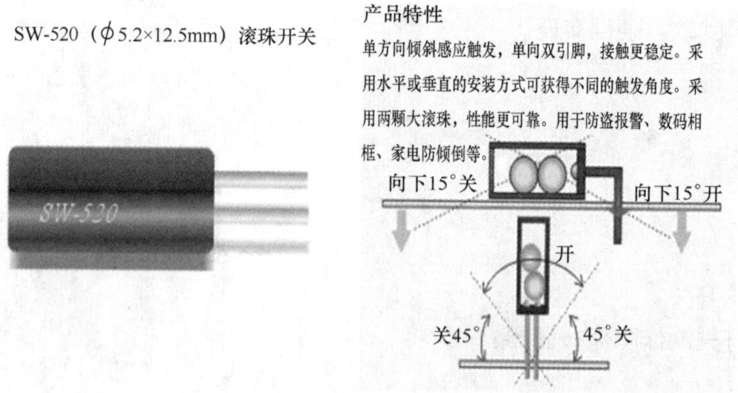
SW-520（ϕ5.2×12.5mm）滚珠开关
SW-520
产品特性
单方向倾斜感应触发，单向双引脚，接触更稳定。采用水平或垂直的安装方式可获得不同的触发角度。采用两颗大滚珠，性能更可靠。用于防盗报警、数码相框、家电防倾倒等。
向下15°关
向下15°开
开
关45°
45°关

图 17–1　滚珠开关实物及原理图

滚珠开关实验程序如下。

```
/**********************BST-M51 实验开发板例程 ************************
*  平台：BST-M51 + Keil U4 + STC89C51
*  名称：滚珠开关实验
*  实验效果：传感器倾斜时，灯灭；正立时，灯亮
************************************************/
#include<reg52.h>

sbit led1 = P1^0;//RED

sbit switch0 = P2^1;

void main()
{
   P1 = 0xff;// 熄灭所有 LED
   while (1)
```

```
    {
    led1 = switch0;
    }
}
```

注意，滚珠开关实验实物连接图如图 17-2 所示，务必按图连接开发板后，再进行整个操作。

图 17-2 滚珠开关实验实物连接图

17.3 项目实施

防盗报警器实验程序如下。

```
/***************BST-M51 实验开发板例程 *************
 * 平台：BST-M51 + Keil U4 + STC89C51
 * 名称：防盗报警实验
 * 实验效果：拾起传感器时，发出蜂鸣声报警，并使 LED 闪烁
 ***********************************************/
#include<reg52.h>

sbit switch0 = P2^1;
sbit beep=P2^3;

void delay()              // 延时 0.5ms 左右
{
   unsigned char a;
   for(a=450;a>0;a--)
   {
   }
}
void main()
{
   P1 = 0xff;
   while(1)
   {
      if(!switch0)// 蜂鸣器及 LED 报警
      {
```

```
            unsigned int m;
            for (m=800;m>0;m--)    //持续时间 0.5ms×800
            {
                beep=~beep;
                delay();            //2kHz 的信号。
            }
            P1=~P1;
            for (m=500;m>0;m--)  //持续时间 0.5ms×2×500
            {
                beep=~beep;
                delay();
                delay();            //1kHz 的信号
            }
            P1=~P1;
        }
    }
}
```

此项目附操作视频及代码资料。

Chapter 18

项目18
8×8点阵显示"爱心"

项目目标

- 8×8LED点阵以LED为像素,分为行控制和列控制,通过单片机的2个端口的引脚输出高电平与低电平来控制二极管的发光来显示文字,注意控制延时来获得较好的视觉效果。

建议学时

- 4学时。

知识要点

- 8×8点阵模块。
- 字模软件。

技能掌握

- 了解汉字的点阵显示原理,运用Proteus软件进行原理图绘制。

18.1 项目分析

通过单片机的控制，使点阵完成一系列的图形的显示与变化，比如静态显示汉字，字母以及数字等，也可在点阵上滚动显示字符。

18.2 技术准备

18.2.1 8×8点阵介绍

8×8点阵共由64个LED组成，且每个LED是放置在行线和列线的交叉点上。引脚号识别方法为点阵模块有4条边，其中一条边上有丝印，且中间向下有塑料突起；插针向下放置，丝印面向使用者，则左起为引脚1，顺次按逆时针排列，如图18-1所示。

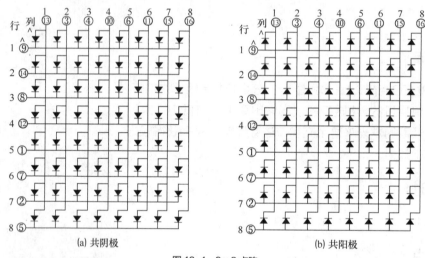

图18-1 8×8点阵

18.2.2 MAX7219介绍

MAX7219/MAX7221是一种集成化的串行输入/输出、共阴极显示驱动器，连接微处理器与8位数字的7段LED数字显示，也可以连接条线图显示器或者64个独立的LED。其上包括1个片上的B型BCD编码器、多路扫描回路、段字驱动器，还有1个8×8的静态RAM，用来存储每一个数据，另外，只有1个外部寄存器用来设置各个LED的段电流。

MAX7219和单片计算机之间有3条引线（DIN，CLK，LOAD）连接，采用16位数据串行移位接收方式，即单片机将16位二进制数逐位发送到DIN端，在CLK上升沿到来前准备就绪。CLK的每个上升沿将1位数据移入MAX7219内移位寄存器，当16位数据移入完毕，在LOAD端口引脚信号上升沿，将16位数据装入MAX7219内的相应位置，在MAX7219内部硬件动态扫描显示控制电路作用下实现动态显示。

18.2.3 MAX7219引脚说明

MAX7219为24引脚芯片，引脚排列如图18-2所示，各引脚功能如下。

DIN：串行数据输入端。

DIG0 ~ DIG7：LED 位线。
LOAD/CS：数据装载信号输入端。
SEGA ~ SEGG, SEGDp：段码输出端。
ISET：硬件亮度调节端。
DOUT：串行数据输出端。
CLK：移位脉冲输入端。
V+：正电源。
GND：地。

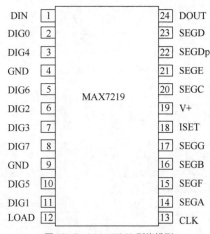

图 18-2　MAX7219 引脚排列

18.2.4　串行数据格式

16 位地址 / 数据移位寄存器接收串行数据，实现串 / 并变换。16 位数据含义如下。

D7 ~ D0：写入内部 RAM 和功能寄存器的数据。

D8 ~ D11：内部 RAM 和功能寄存器地址。

D12 ~ D15：无定义。

地址译码器是一个 4 ~ 16 线译码器，用于选择数据存放单元，在 LOAD 端口信号作用下，将接收数据送入指定单元；8 字节双端口静态存储器存放接收数据和提供动态显示数据；B 译码和不译码电路对 RAM 数据进行 BCD 译码或直接送显示；段码电流参考电路、亮度脉冲产生调制器实现对显示器的亮度控制，段码电流参考电路由硬件调节显示器亮度；动态扫描控制器实现由硬件控制动态扫描显示。LED 段 / 位驱动器提供显示器的 1 段和 1 位点亮时的电流。

18.2.5　可寻址的数据寄存器和控制寄存器

（1）内部 RAM 地址 01 ~ 08H 分别对应于 DIG0 ~ DIG7。

（2）译码方式寄存器（地址 09H）：该寄存器的 8 位二进制数的各位分别控制 8 个 LED 显示器的译码方式。当高电平时，选择 BCD-B 译码模式，当低电平时选择不译码模式（即送来数据为字型码）。

（3）亮度寄存器（地址 0AH）：亮度可以用硬件和软件 2 种方法调节。亮度寄存器中的 D0 ~ D3 位可以控制 LED 显示器的亮度。

（4）扫描界限寄存器（地址 0BH）：该寄存器中 D0 ~ D3 位数据设定值为 0 ~ 7H，设定值表示显示器动态扫描个数位 1 ~ 8。

（5）停机寄存器（地址 0CH）：当 D0=0 时，MAX721 处于停机状态；当 D0=1 时，MAX7219 处于正常工作状态。

（6）显示测试寄存器（地址 0FH）：当 D0=0 时，MAX7219 按设定模式正常工作；D0=1 时，MAX7219 处于测试状态。在该状态下，不管 MAX7219 处于什么模式，全部 LED 将按最大亮度显示。

MAX7219 写入字节程序如下。

```
//------------------------------------------
// 功能：向 MAX7219（U3）写入字节
// 入口参数：DATA
// 出口参数：无
// 说明：
void Write_Max7219_byte(uchar DATA)
{
    uchar i;
        Max7219_pinCS=0;
        for (i=8;i>=1;i--)
```

```
            {
                Max7219_pinCLK=0;
                Max7219_pinDIN=DATA&0x80;
                DATA=DATA<<1;
                Max7219_pinCLK=1;
            }
}
```

MAX7219 写入数据程序如下。

```
//------------------------------------------
// 功能：向 MAX7219 写入数据
// 入口参数：address、dat
// 出口参数：无
void Write_Max7219 (uchar address,uchar dat)
{
    Max7219_pinCS=0;
    Write_Max7219_byte (address);          // 写入地址，即 LED 显示器编号
    Write_Max7219_byte (dat);              // 写入数据，即 LED 显示器显示数字
    Max7219_pinCS=1;
}
```

MAX7219 初始化程序如下。

```
void Init_MAX7219 (void)
{
    Write_Max7219 (0x09, 0x00);     // 译码方式为 BCD 码
    Write_Max7219 (0x0a, 0x03);     // 亮度
    Write_Max7219 (0x0b, 0x07);     // 扫描界限；8 个 LED 显示器显示
    Write_Max7219 (0x0c, 0x01);     // 掉电模式为 0，普通模式为 1
    Write_Max7219 (0x0f, 0x00);     // 显示测试为 1；测试结束，正常显示为 0
}
```

18.3 项目实施

8×8 点阵模块显示爱心形状实验程序如下。

```
/********************BST-M51 实验开发板例程 ************************
*  平台：BST-M51 + Keil U4 + STC89C51
*  名称：8×8 点阵模块实验
*  晶振：11.059 2MHz
*  实验效果：显示爱心形状
*****************************************************************/
#include <reg52.h>
#include <intrins.h>
#define uchar unsigned char
#define uint  unsigned int
// 定义 MAX7219 端口
sbit Max7219_pinCLK = P2^2;//CLK
sbit Max7219_pinCS  = P2^1;//CS
sbit Max7219_pinDIN = P2^0;//DIN
uchar code disp1[8]=
{0x66,0x99,0x81,0x81,0x42,0x24,0x18,0x00};// "爱心"

void Delay_xms (uint x)
```

```c
{
    uint i,j;
    for(i=0;i<x;i++)
     for(j=0;j<112;j++);
}
//--------------------------------------
// 功能：向MAX7219（U3）写入字节
// 入口参数：DATA
// 出口参数：无
// 说明：
void Write_Max7219_byte(uchar DATA)
{
    uchar i;
        Max7219_pinCS=0;
      for(i=8;i>=1;i--)
            {
              Max7219_pinCLK=0;
              Max7219_pinDIN=DATA&0x80;
              DATA=DATA<<1;
              Max7219_pinCLK=1;
            }
}
//--------------------------------------
// 功能：向MAX7219写入数据
// 入口参数：address,dat
// 出口参数：无
void Write_Max7219(uchar address,uchar dat)
{
    Max7219_pinCS=0;
    Write_Max7219_byte(address);      // 写入地址，即LED显示器编号
      Write_Max7219_byte(dat);        // 写入数据，即LED显示器显示数字
    Max7219_pinCS=1;
}

void Init_MAX7219(void)
{
    Write_Max7219(0x09, 0x00);        // 译码方式为BCD码
    Write_Max7219(0x0a, 0x03);        // 亮度
    Write_Max7219(0x0b, 0x07);        // 扫描界限；8个LED显示器显示
    Write_Max7219(0x0c, 0x01);        // 掉电模式为0；普通模式为1
    Write_Max7219(0x0f, 0x00);        // 显示测试为1；测试结束，正常显示为0
}
void main(void)
{
    uchar i;
    Delay_xms(50);
    Init_MAX7219();
    for(i=1;i<9;i++)
        Write_Max7219(i,disp1[i-1]);
}
```

注意，显示爱心形状实验实物连接图如图18-3所示，务必按图连接开发板后，再进行整个操作。

图 18-3　显示爱心形状实验实物连接图

18.4　技术拓展

8×8 点阵常用显示实验程序如下。

```
/********************BST-M51 实验开发板例程 ***********************
*   平台：BST-M51 + Keil U4 + STC89C51
*   名称：8×8 点阵模块实验
*   晶振：11.059 2MHz
*   实验效果：显示 0～9、A～Z、中、国
***************************************************************/
#include <reg52.h>
#include <intrins.h>
#define uchar unsigned char
#define uint  unsigned int
// 定义 Max7219 端口
sbit Max7219_pinCLK = P2^2;//CLK
sbit Max7219_pinCS  = P2^1;//CS
sbit Max7219_pinDIN = P2^0;//DIN
uchar code disp1[38][8]={
{0x3C,0x42,0x42,0x42,0x42,0x42,0x42,0x3C},//0
{0x10,0x18,0x14,0x10,0x10,0x10,0x10,0x10},//1

// 上面列举了"数字 0,1"的显示。在附带文件里面有字符代码生成器，也有详细视频教程，可动手自行生成其他有创意的文字。

};
void Delay_xms (uint x)
{
  uint i,j;
  for (i=0;i<x;i++)
    for (j=0;j<112;j++);
}
//--------------------------------------------
// 功能：向 MAX7219（U3）写入字节
```

```c
// 入口参数: DATA
// 出口参数: 无
void Write_Max7219_byte (uchar DATA)
{
   uchar i;
      Max7219_pinCS=0;
      for (i=8;i>=1;i--)
          {
           Max7219_pinCLK=0;
           Max7219_pinDIN=DATA&0x80;
           DATA=DATA<<1;
           Max7219_pinCLK=1;
          }
}
//------------------------------------------
// 功能：向 MAX7219 写入数据
// 入口参数: address, dat
// 出口参数: 无
void Write_Max7219 (uchar address,uchar dat)
{
    Max7219_pinCS=0;
  Write_Max7219_byte (address);    // 写入地址，即 LED 显示器编号
    Write_Max7219_byte (dat);      // 写入数据，即 LED 显示器显示数字
   Max7219_pinCS=1;
}

void Init_MAX7219 (void)
{
  Write_Max7219 (0x09, 0x00);    // 译码方式为 BCD 码
  Write_Max7219 (0x0a, 0x03);    // 亮度
  Write_Max7219 (0x0b, 0x07);    // 扫描界限；8 个 LED 显示器显示
  Write_Max7219 (0x0c, 0x01);    // 掉电模式为 0，普通模式为 1
  Write_Max7219 (0x0f, 0x00);    // 显示测试为 1；测试结束，正常显示为 0
}
void main (void)
{
  uchar i,j;
  Delay_xms (50);
  Init_MAX7219();
  while (1)
   {
   for (j=0;j<38;j++)
    {
    for (i=1;i<9;i++)
      Write_Max7219 (i,disp1[j][i-1]);
Delay_xms (1000);
   }
  }
}
```

此项目附操作视频和字符代码生成器。

Chapter 19

项目19
温度计显示

项目目标

- 主机控制DS18B20完成温度转换必须经过3个步骤，即初始化、ROM操作指令、存储器操作指令。必须先启动DS18B20开始转换，再读出温度转换值。本程序仅挂接1个芯片，使用默认的12位转换精度，外接供电电源，读取的温度值高位字节送WDMSB单元，低位字节送WDLSB单元，再按照温度值字节的表示格式及其符号位，经过简单的变换即可得到实际温度值。

建议学时

- 4学时。

知识要点

- DS 18B20相关编程。
- 初步认识DS 18B20温度显示器件。

技能掌握

- 采用数字式温度传感器为检测器件，进行单点温度检测。用LCD直接显示温度值。

19.1 项目分析

DS18B20 是 DALLAS 公司生产的一线式数字温度传感器,具有 3 引脚 TO-92 小体积封装形式;温度测量范围为 -55℃ ~ +125℃,可编程为 9 ~ 12 位 A/D 转换精度,测温分辨率可达 0.062 5℃。

19.2 技术准备

19.2.1 DS18B20 单线总线的工作方式

1. DS18B20 简介

(1) DS18B20 单线数字温度传感器,即"一线器件",其具有如下独特的优点。

① 采用单线总线的接口方式。与微处理器连接时,仅需要 1 条口线即可实现微处理器与 DS18B20 的双向通信。单线总线具有经济性好、抗干扰能力强、适合于恶劣环境的现场温度测量、使用方便等优点,使用户可轻松组建传感器网络,为测量系统的构建引入全新概念。

② 测量温度范围宽,测量精度高。DS18B20 的测量范围为 -55℃ ~ +125℃;在 -10 ~ +85℃范围内,精度为 ±0.5℃。

③ 支持多点组网功能。多个 DS18B20 可以并联在单线上,实现多点测温。

④ 供电方式灵活。DS18B20 可以通过内部寄生电路从数据线上获取电源。因此,当数据线上的时序满足一定的要求时,可以不接外部电源,从而使系统结构更趋简单,可靠性更高。

⑤ 测量参数可配置。DS18B20 的测量分辨率可通过程序设定 9 ~ 12 位。

DS18B20 具有体积更小、更经济、更宽的电压适用范围,可选更小的封装方式,适合于构建经济型测温系统,因而被设计者们所青睐。

(2) 单线总线特点如下。

单线总线即只有 1 根数据线,系统中的数据交换、控制都由这根线完成。单线总线通常要求外接 1 个 4.7 ~ 10kΩ 的上拉电阻,这样,当总线闲置时其状态为高电平。DS18B20 原理图如图 19-1 所示。

2. DS18B20 的初始化

主机首先发出一个 480 ~ 960μs 的低电平脉冲,然后释放总线变为高电平,并在随后的 480μs 时间内对总线进行检测,如果有低电平出现,说明总线上有器件已做出应答;若无低电平出现,一直都是高电平,说明总线上无器件应答。

从 DS18B20 通电开始,就一直在检测总线上是否有 480 ~ 960μs 的低电平出现,如果有,在总线转为高电平后,等待 15 ~ 60μs 后,将总线电平拉低 60 ~ 240μs,做为存在响应的脉冲,通知主机本器件已做好准备;若没有检测到,就一直检测等待。DS1820 初始化时序图如图 19-2 所示。

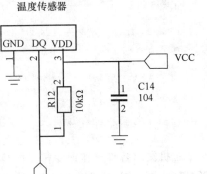

图 19-1 DS18B20 原理图

初始化过程"复位和存在脉冲"

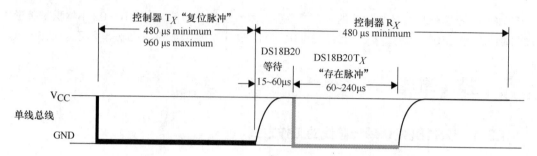

图 19-2　DS1820 初始化时序图

DS18B20 初始化程序如下。

```
/***************************************************************
* 函数名：DS18B20Init
* 函数功能：初始化
* 输入：无
* 输出：初始化成功，返回 1；失败，返回 0
***************************************************************/
unsigned char Ds18b20Init()
{
    unsigned int i;
    DSIO=0;                    //将总线电平拉低 480～960μs
    i=70;
    while(i--);//延时 642μs
    DSIO=1;//然后拉高总线电平，若 DS18B20 做出反应会将在 15～60μs 后将总线电平拉低
    i=0;
    while(DSIO)//等待 DS18B20 拉低总线电平
    {
        i++;
        if(i>50000)//等待时间 >50ms
            return 0;//初始化失败
    }
    return 1;//初始化成功
}
```

主机发出各种操作命令都是向 DS18B20 写"0"和"1"组成的命令字节，接收数据时也是从 DS18B20 读取"0"或"1"的过程。因此首先要掌握主机是如何进行写"0"、写"1"、读"0"和读"1"的。

写周期最少为 60μs，最长不超过 120μs。主机先把总线电平拉低 1μs，表示写周期开始；随后若主机写"0"，则将总线置为低电平，若主机写"1"，则将总线置为高电平，持续时间最少 60μs，直至写周期结束；然后释放总线为高电平至少 1μs，使总线恢复。而 DS18B20 则在检测到总线电平被拉低后等待 15μs，然后在 15～45μs 开始对总线电平采样，在采样期内总线为高电平，则为"1"，若采样期内总线为低电平，则为"0"。写操作时序图如图 19-3 所示。

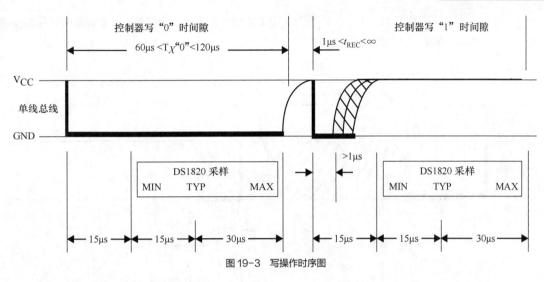

图 19-3 写操作时序图

向 DS18B20 写入 1 字节程序如下。

```
/************************************************************
* 函数名：DS18B20WriteByte
* 函数功能：向 DS18B20 写入 1 字节        * 输入：dat        * 输出：无
************************************************************/
void tmpwritebyte(uchar dat)   // 向 DS18B20 写入字节
{
  uint i;
  uchar j;
  bit testb;                    // 定义一个位变量
  for(j=1;j<=8;j++)             //1 字节 8 位数据，每次只能写 1 位
  {
    testb=dat&0x01;             // 依次将 dat 的每一位赋值给 testb
    dat=dat>>1;
    if(testb)    //write 1
    {
      DS=0;
      i++;i++;                  // > 1μs
      DS=1;
      i=8;while(i>0) i--;       // ≥ 60μs
    }
    else
    {
      DS=0;         //write 0
      i=8;while(i>0) i--;  // 至≥ 60μs
      DS=1;
      i++;i++;
    }
  }
}
```

对于读数据操作时序，也分为读"0"时序和读"1"时序 2 个过程。

读周期一开始，主机把总线电平拉低 1μs 之后，就得释放总线为高电平，以让 DS18B20 把数据传输到总线上。作为从机，DS18B20 在检测到总线被拉低 1μs 后，便开始送出数据。若是要送出"0"，就把总线电平拉低，直到读周期结束；若要送出"1"，则释放总线为高电平。主机在一开始拉低总线电平 1μs 后释放总线，然后在包括此拉低总线电平 1μs 在内的 15μs 时间内，完成对总线的采样检测。采

样期内总线为低电平，则确认为"0"；采样期内总线为高电平，则确认为"1"。1个读时序过程至少需要60μs才能完成。该操作时序图如图19-4所示。

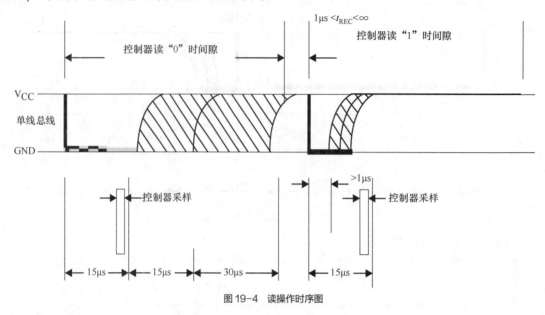

图19-4 读操作时序图

DS18B20读取1位数据的程序如下。

```
/***************************************************************
* 函数名：DS18B20ReadByte
* 函数功能：读取1位数据
* 输入 : com
* 输出         : 无
***************************************************************/
bit Ds18b20ReadByte(void)       //read a bit
{
  uint i;
  bit dat;
  DS=0;i++;              //i++ for delay
  DS=1;i++;i++;
  dat=DS;
  i=8;while(i>0) i--;
  return (dat);
}
```

19.2.2 DS18B20的操作步骤

DS18B20的一线工作协议流程为初始化→ROM操作指令→存储器操作指令→数据传输。其工作时序包括初始化时序、写时序、读时序。

DS18B20内部结构主要由四部分组成，即64位光刻ROM、温度传感器、非挥发的温度报警触发器TH和TL、配置寄存器。

光刻ROM中的64位序列号是出厂前被光刻好的，它可以看作是该DS18B20的地址序列码，见表19-1。64位光刻ROM的排列是，开始8位（地址为28H）是产品类型标号；接着的48位是该DS18B20自身的序列号，并且每个DS18B20的序列号都不相同，因此可以看做是该DS18B20的地址序列码；最后8位则是前面56位的循环冗余校验码（CRC=X8+X5+X4+1）。由于每一个DS18B20的ROM数据都各不相同，因此微控制器就可以通过单线总线对多个DS18B20进行寻址，从而实现1根总线上挂接多个 DS18B20 的目的。

其中配置寄存器的格式见表 19-1，与分辨率的关系见表 19-2。

表 19-1 配置寄存器的格式

0	R1	R0	1	1	1	1	1
MSB							LSB

表 19-2 配置寄存器与分辨率关系

R0	R1	温度计分辨率 /bit	最大转换时间 /ms
0	0	9	93.75
0	1	10	187.75
1	0	11	375
1	1	12	750

DS18B20 经转换所得的温度值，以 2 字节补码形式存放在高速暂存存储器的第 0 和第 1 个字节。所以当只想简单的读取温度值的时候，只用读取暂存器中的第 0 和第 1 个字节就可以了。

简单的读取温度值的步骤如下。

（1）跳过 ROM 操作。

（2）发送温度转换命令。

（3）跳过 ROM 操作。

（4）发送读取温度命令。

（5）读取温度值。

正数的补码是正数本身；负数的补码是原码取反，然后再加 1。

DS18B20 存储的温度值是以补码的形式存储的，所以读出来的温度值是实际温度值的补码，要把补码转换为原码。

正温度的原码就是补码本身，所以在 12 位分辨率下，温度的计算公式是为

$$温度值 = 读取值 \times 0.062\ 5$$

负温度的原码是补码减 1，再取反，所以在 12 位分辨率下，计算公式为

$$温度值 = -（读取值减 1 再取反）\times 0.062\ 5$$

DS18B20 的指令集合见表 19-3。

表 19-3 DS18B20 的指令合集

指 令 名 称	指 令 代 码	指 令 功 能
温度变换	44H	启动 DS18B20 进行温度转换，转换时间最长为 500ms（典型为 200ms），结果存入内部 9 字节 RAM 中
读暂存器	0BEH	读内部 RAM 中 9 字节的内容
写暂存器	4EH	发出向内部 RAM 的第 3，4 字节写上，下限温度数据命令，紧跟该命令之后，是传送两字节的数据
复制暂存器	48H	将 RAM 中第 3，4 字节的内容复制到 EEPROM 中
重调 EEPROM	0B8H	EEPROM 中的内容恢复到 RAM 中的第 3，4 字节
读供电方式	0B4H	读 DS18B20 的供电模式，寄生供电时 DS18B20 发送"0"，外接电源供电 DS18B20 发送"1"

续表

指令名称	指令代码	指令功能
读 ROM	33H	读 DS18B20ROM 中的编码（即读 64 位地址）
ROM 匹配（符合 ROM）	55H	发出此命令之后，接着发出 64 位 ROM 编码，访问单总线上与编码相对应 DS18B20 使之作出响应，为下一步对该 DS18B20 的读写作准备
搜索 ROM	0F0H	用于确定挂接在同一总线上 DS18B20 的个数和识别 64 位 ROM 地址，为操作各器件做好准备
跳过 ROM	0CCH	忽略 64 位 ROM 地址，直接向 DS18B20 发温度变换命令，适用于单片机工作
警报搜索	0ECH	该指令执行后，只有温度超过设定值上限或下限的片子才做出响应

```c
/*****************************************************************
* 函数名 : DS18B20ReadTemp
* 函数功能：读取温度
* 输入 : com
* 输出：无
*****************************************************************/
int Ds18b20ReadTemp()
{
    unsigned int temp=0;
    unsigned char tmh,tml;
    Ds18b20ChangTemp();        // 先写入转换命令
    Ds18b20ReadTempCom();      // 然后等待转换完毕后，发送读取温度命令
    tml=Ds18b20ReadByte();     // 读取温度值共 16 位，先读低字节
    tmh=Ds18b20ReadByte();     // 再读高字节
    temp=tmh;
    temp<<=8;
    temp|=tml;
    return temp;
}


/*********************************************
* 函数名: DS18B20ChangTemp
* 函数功能   : 让 DS18B20 开始转换温度
* 输入 : com
* 输出     : 无
*********************************************/
void Ds18b20ChangTemp()
{
    Ds18b20Init();
    Delay1ms(1);
    Ds18b20WriteByte(0xcc);    // 跳过 ROM 操作命令
    Ds18b20WriteByte(0x44);    // 温度转换命令
    Delay1ms(100);
}


/*********************************************
* 函数名 : DS18B20ReadTempCom
```

* 函数功能：发送读取温度命令
* 输入 ：com
* 输出   ：无
**/
```c
void  Ds18b20ReadTempCom()
{
   Ds18b20Init();
   Delay1ms(1);
   Ds18b20WriteByte(0xcc);    // 跳过ROM操作命令
   Ds18b20WriteByte(0xbe);    // 发送读取温度命令
}
```

19.3 项目实施

DS18B20 LED 显示器温度显示实验程序如下。

```c
/********************BST-M51实验开发板例程 ***********************
*  平台：BST-M51 + Keil U4 + STC89C51
*  名称：DS18b20温度显示（LED显示器）实验
*  晶振：11.059 2MHz
*  实验效果：接上DS18B20温度传感器LED显示器显示出当前温度
**************************************************************/
#include <reg52.h>
#define uchar unsigned char
#define uint unsigned int
#define dula P0                  // 段选信号的锁存器控制
sbit wei1=P2^4;
sbit wei2=P2^5;
sbit wei3=P2^6;
sbit wei4=P2^7;
sbit DS=P2^2;                   // 定义接口
uint temp;                      // 变量
unsigned char code table[]={0x3f,0x06,0x5b,0x4f,0x66,0x6d,0x7d,
                    0x07,0x7f,0x6f,0x77,0x7c,0x39,0x5e,0x79,0x71};
unsigned char code table1[]={0xbf,0x86,0xdb,0xcf,0xe6,0xed,0xfd,
                    0x87,0xff,0xef};

void delay(uint count)          // 延迟
{
  uint i;
  while(count)
  {
    i=200;
    while(i>0)
    i--;
    count--;
  }
}


void dsreset(void)              // 发送复位和初始化命令
{
   uint i;
   DS=0;
```

```c
    i=103;              //将总线拉电平拉低480～960μs
    while(i>0) i--;
    DS=1;               //然后拉高总线电平,若DS18B20做出反应,将会在15～60μs后,将总线电平拉低
    i=4;                //15～60μs等待
    while(i>0) i--;
    //while(DS);
}

bit tmpreadbit(void)
{
    uint i;
    bit dat;
    DS=0;i++;           //i++ 延迟
    DS=1;i++;i++;
    dat=DS;
    i=8;while(i>0) i--;
    return(dat);
}

uchar tmpread(void)   //一个字节的延迟
{
    uchar i,j,dat;
    dat=0;
    for(i=1;i<=8;i++)
    {
        j=tmpreadbit();
dat=(j<<7)|(dat>>1);    //读出的数据最低位在最前面,这样刚好1字节在dat里
    }
    return(dat);
}

void tmpwritebyte(uchar dat)   //写一个字节的DS18B20
{
    uint i;
    uchar j;
    bit testb;
    for(j=1;j<=8;j++)
    {
        testb=dat&0x01;
        dat=dat>>1;
        if(testb)    //write 1
        {
            DS=0;
            i++;i++;
            DS=1;
            i=8;while(i>0) i--;
        }
        else
        {
            DS=0;        //write 0
            i=8;while(i>0) i--;
            DS=1;
            i++;i++;
        }
    }
}
```

```c
}

void tmpchange(void)           //DS18B20 开始改变
{
  dsreset();
  delay(1);
  tmpwritebyte(0xcc);
  tmpwritebyte(0x44);          //  启动一个单一的温度转换
  //delay(100);
}

uint tmp()                     // 得到温度
{
  float tt;
  uchar a,b;
  dsreset();
  delay(1);
  tmpwritebyte(0xcc);
  tmpwritebyte(0xbe);
  a=tmpread();// 低八位
  b=tmpread();// 高八位
  temp=b;
  temp<<=8;                    // 两个字节组成一个整型变量
  temp=temp|a;
  tt=temp*0.0625;              // 算出来的是测到的温度，数值可到小数点后两位
temp=tt*10+0.5;                // 为了显示温度后的小数点后一位，并作出四舍五入，因为取值运算不能取小数点后的数
  return temp;
}

void display(uint temp)        // 显示程序
{
   uchar bai,shi1,shi0,ge;
   bai=temp/100;// 温度数值上为十位
   shi0=temp%100;// 温度数值上为几点几
   shi1=shi0/10;// 温度上为个位,并且显示时需要加小数点
   ge=shi0%10;// 温度上为小数位,并已经四舍五入

   wei1=1;              // 显示百位
   wei2=0;
   wei3=0;
   wei4=0;
   P0=table[bai];
   delay(2);

   wei1=0;              // 显示十位
   wei2=1;
   wei3=0;
   wei4=0;
   P0=table1[shi1];
   delay(2);

   wei1=0;              // 显示个位
   wei2=0;
   wei3=1;
   wei4=0;
   P0=table[ge];
```

```c
    delay(2);
}
void main()
{
  uchar a;
do
  {
    tmpchange();//让DS18B20开始转换温度

  for(a=100;a>0;a--)
  {
    display(tmp());
  }
  }while(1);
}
```

19.4 技术拓展

DS18B20 LCD1602 温度显示实验程序如下。

```c
/********************BST-M51 实验开发板例程 ********************
*  平台: BST-M51 + Keil U4 + STC89C51
*  名称: DS18b20温度显示（LCD1602）实验
*  晶振: 11.059 2MHz
*  实验效果: 接上DS18B20温度传感器,LCD1602显示出当前温度
****************************************************************/
#include <reg52.H>
#include <intrins.H>
#include <math.H>
#define uchar unsigned char
  #define uint unsigned int
  sbit dula = P2^6;
  sbit wela = P2^7;
  sbit rw = P1^1;
  sbit RS = P1^0;
sbit LCDEN = P2^5;
void delayUs()
{
    _nop_();
}

  void delayMs(uint a)
{
    uint i, j;
    for(i = a; i > 0; i--)
        for(j = 100; j > 0; j--);
    }

/************************LCD1602*************************/
void writeComm(uchar comm)
{
    RS = 0;
    P0 = comm;
    LCDEN = 1;
```

```c
        delayUs();
        LCDEN = 0;
        delayMs(1);
}

// 写数据:RS=1, RW=0;
void writeData(uchar dat)
{
    RS = 1;
    P0 = dat;
    LCDEN = 1;
    delayUs();
    LCDEN = 0;
    delayMs(1);
}
void init()
{
    rw = 0;
    dula = wela = 0;
    writeComm(0x38);
    writeComm(0x0c);
    writeComm(0x06);
    writeComm(0x01);
}

void writeString(uchar * str, uchar length)
{
    uchar i;
    for (i = 0; i < length; i++)
    {
        writeData(str[i]);
    }
}

/*************************DS18B20****************************/
参考附带视频
/**//*********************** 主函数 ****************************/
void main()
{
    uchar table[] = "  xianzaiwendu: ";
    sendChangeCmd();
    init();
    writeComm(0x80);
    writeString(table, 16);
    while (1)
{
        delayMs(1000); // 温度转换时间需要750ms以上
writeComm(0xc0);
        display(getTmpValue());
        sendChangeCmd();
}
}
```

此项目附操作视频及代码资料。

20 Chapter

项目20
测距显示

项目目标

- 通过构造单片机开发环境,了解单片机开发系统结构和流程。

建议学时

- 4学时。

知识要点

- 使用超声波传感器编程实现功能。
- 超声波传感器的基本结构。

技能掌握

- 了解超声波传感器测距的原理,掌握使用LCD显示距离。

20.1 项目分析

测距显示使用的是超声波传感器，超声波传感器的主要部件是超声波探头。超声波探头主要由压电晶片组成，既可以发射超声波，也可以接收超声波。小功率超声探头多作探测用。按结构的不同，可分为直探头（纵波）、斜探头（横波）、表面波探头（表面波）、兰姆波探头（兰姆波）、双探头（一个探头发射、一个探头接收）等。

20.2 技术准备

20.2.1 HC-SR04 超声波测距模块

HC-SR04 超声波测距模块实物如图 20-1 所示。

图 20-1 HC-SR04 超声波测距模块

1. HC-SR04 超声波测距模块产品特点

（1）典型工作用电压：5V。
（2）超小静态工作电流：< 2mA。
（3）感应角度：≤ 15°。
（4）探测距离：2 ~ 400cm。
（5）高精度：可达 0.3cm。
（6）盲区：2cm。

2. HC-SR04 超声波测距模块端口及使用方法

HC-SR04 超声波测距模块有 4 个端口，如图 20-1 所示，分别为 Vcc（电源）、Trig（控制端口）、Echo（接收端口）、Gnd（接地）。

本产品使用方法：控制端口发了出一个 10μs 以上的高电平，就可以在接收端口等待高电平输出。一有输出，就可以开定时器计时，当此端口变为低电平时，就可以读定时器的值，此值即为本次测距的时间，据此即可算出所测距离。HC-SR04 电气参数见表 20-1。

表 20-1　HC-SR04 超声波测距模块电气参数

电 气 参 数	HC-SR04 超声波测距模块
工作电压	DC5V
工作电流	15mA
工作频率	40Hz
最远射程	4m
最近射程	2cm
测量角度	15°
输入触发信号	10μs 的 TTL 脉冲
输出回响信号	输出 TTL 电平信号，与射程成比例
规格尺寸	45×20×15mm

20.2.2　超声波测距原理

1．超声波的概念

超声波是一种频率比较高的声音，指向性强。超声波测距的原理是利用超声波在空气中的传播速为已知，测量声波在发射后遇到障碍物反射回来的时间，根据发射和接收的时间差，计算出发射点到障碍物的实际距离。由此可见，超声波测距原理与雷达原理是一样的。

测距的公式为

$$L = C \times T$$

式中，L 为测量的距离长度；C 为超声波在空气中的传播速度（已知 20℃室温时 C=344m/s）；T 为测量距离传播的时间差（T 为发射到接收时间数值的一半）。

现实中，超声波传播速度存在误差。超声波的传播速度会受到空气密度的影响，空气的密度越高，则超声波的传播速度越快，而空气的密度又与温度有着密切的关系。因此超声波传播速度实际近似计算公式为

$$C = C_0 + 0.607 \times T$$

式中，C_0 为零度时的声波速度 332m/s；T 为实际环境温度（℃）。

对于超声波测距精度要求达到 1mm 时，就必须把超声波传播的环境温度考虑进去。

2．超声波模块工作原理

（1）采用 I/O 触发测距，给至少 10μs 的高电平信号。

（2）模块自动发送 8 个 40kHz 的方波，自动检测是否有信号返回。

（3）有信号返回，通过 I/O 输出高电平，高电平持续时间就是超声波从发射到返回的时间。

（4）测试距离 =[高电平时间 × 声速（340m/s）]/2。

超声波测距时序图如图 20-2 所示，由图可见，只要提供 1 个 10μs 以上的脉冲触发信号，该模块内部就会发出 8 个 40kHz 周期电平，并检测回波。一旦检测到有回波信号，则输出回响信号，由回响信号时间间隔即可计算出距离。建议测量周期为 60ms 以上，以防止发射信号对回响信号的影响。超声波测距还会常用到 2 个换算公式，即 μs/58= 厘米或者 μs/148= 英寸。

开发板超声波测距模块电路原理图如图 20-3 所示。

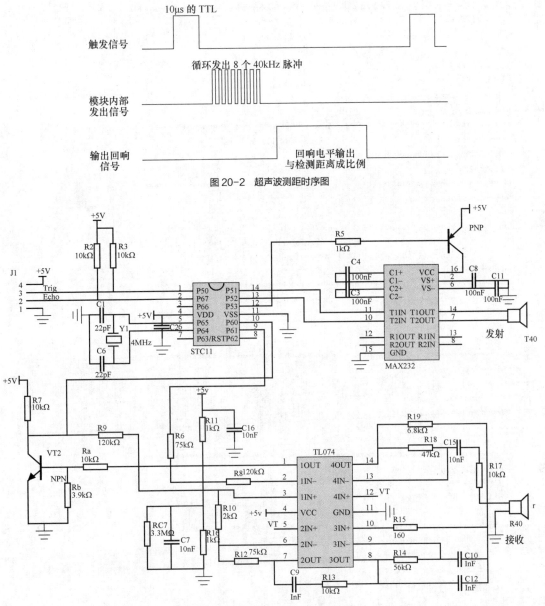

图 20-2 超声测距时序图

图 20-3 开发板超声波测距模块电路原理图

20.3 项目实施

超声波测距 LED 显示器实验程序如下。

```
/*********************BST-M51 实验开发板例程 *********************
*  平台：BST-M51 + Keil U4 + STC89C51
*  名称：超声波测距实验
*  晶振：11.059 2MHz
*****************************************************************/
/*************************** 包含头文件 ***************************/
#include <reg52.h>
```

```c
/*************************** 位定义 ***************************/
sbit echo = P2^0;                    // 回声接收端口
sbit trig = P2^1;                    // 超声触发端口

sbit wei1=P2^4;
sbit wei2=P2^5;
sbit wei3=P2^6;
sbit wei4=P2^7;
/*************************** 宏定义 ***************************/
#define dula P0                      // 段选信号的锁存器控制
#define uchar unsigned char
#define uint unsigned int
/*********************** 定义变量和数组 ***********************/
long int distance=0;                 // 距离变量
uchar count;
unsigned char code table[]={0x3f,0x06,0x5b,0x4f,0x66,0x6d,0x7d,
                            0x07,0x7f,0x6f,0x77,0x7c,0x39,0x5e,0x79,0x71};
unsigned char code table1[]={0xbf,0x86,0xdb,0xcf,0xe6,0xed,0xfd,
0x87,0xff,0xef,0x71};
/***************************************************************/
/* 函数名称：delay                                              */
/* 函数描述：延时函数                                           */
/* 输入参数：延时时间（ms）                                     */
/* 返回值 ：无                                                  */
/***************************************************************/
void delay(unsigned int xms)
{
    unsigned int i,j;
    for(i=xms;i>0;i--)        //i=xms 即延时 xms
        for(j=112;j>0;j--);
}
/***************************************************************/
/* 函数名称：display                                            */
/* 函数描述：LED 显示器显示函数                                 */
/* 输入参数：测试距离值的 10 倍                                 */
/* 返回值：无                                                   */
/***************************************************************/
void display(long int num)           // 显示程序
{
    uchar qian,bai,shi,ge;
    if((num>5000) || (num == 0))
    {
        qian=15;// 距离值上为百位
        bai=15;
        shi=10;
        ge=15;
    }
    else
    {
        qian=num/1000;// 距离值上为百位
        bai=(num/100)%10;// 距离值上为十位
        shi=(num/10)%10;// 距离值上为个位,并且显示时需要加小数点
        ge=num%10;// 距离值上为小数位,并已经四舍五入
    }
    wei1=1;                   // 显示千位
    wei2=0;
    wei3=0;
```

```c
    wei4=0;
    P0=table[qian];
    delay(2);

    wei1=0;              // 显示百位
    wei2=1;
    wei3=0;
    wei4=0;
    P0=table[bai];
    delay(2);

    wei1=0;              // 显示十位
    wei2=0;
    wei3=1;
    wei4=0;
    P0=table1[shi];
    delay(2);

    wei1=0;              // 显示个位
    wei2=0;
    wei3=0;
    wei4=1;
    P0=table[ge];
    delay(2);
}
/*************************************************************************/
/* 函数名称：nit_timer2                                                    */
/* 函数描述：初始化单片机函数                                              */
/* 输入参数：无                                                            */
/* 参数描述：无                                                            */
/* 返回值 ：无                                                             */
/*************************************************************************/
void Init_timer0(void)
{
    TMOD = 0x01;         // 定时器 T0 初始化，设置为 16 位自动重装模式
    TL0 = 0x66;
    TH0 = 0xfc;          //1ms
    ET0 = 1;             // 开定时器 T2
    EA = 1;              // 总中断使能
}
/*************************************************************************/
/* 函数名称：Init_Parameter                                                */
/* 函数描述：初始化参数和 I/O 端口函数                                     */
/* 输入参数：无                                                            */
/* 参数描述：无                                                            */
/* 返回值：无                                                              */
/*************************************************************************/
void Init_Parameter(void)
{
    echo = 0;
    trig = 0;
    count = 0;
    distance = 0;
}
/*************************************************************************/
```

```c
/* 函数名称：Trig_SuperSonic                                          */
/* 函数描述：发出声波函数                                             */
/* 输入参数：无                                                       */
/* 参数描述：无                                                       */
/* 返回值：无                                                         */
/*********************************************************************/
void Trig_SuperSonic (void) // 发出声波
{
    trig = 1;
    delay(1);
    trig = 0;
}
/*********************************************************************/
/* 函数名称：Measure_Distance                                         */
/* 函数描述：计算距离函数                                             */
/* 输入参数：无                                                       */
/* 参数描述：无                                                       */
/* 返回值：无                                                         */
/*********************************************************************/
void Measure_Distance (void)
{
    uchar l;
    uint h, y;
    TR0 = 1;
    while (echo)
    {
        ;
    }
    TR0 = 0;
    l = TL0;
    h = TH0;
    y = (h << 8) + l;
    y = y - 0xfc66;// μs 部分
    distance = y + 1000 * count;// 计算总时间（μs）
    TL0 = 0x66;
    TH0 = 0xfc;
    delay(30);
    distance = 0.17 * distance;//
}
/*********************************************************************/
/* 函数名称：main                                                     */
/* 函数描述：主函数                                                   */
/* 输入参数：无                                                       */
/* 参数描述：无                                                       */
/* 返回值：无                                                         */
/*********************************************************************/
void main (void)
{
    uchar a;
    Init_timer0();
    Init_Parameter();
    while (1)
    {
        Trig_SuperSonic();           // 触发超声波发射
        while (echo == 0)            // 等待回声
        {
```

```
              ;
         }
         Measure_Distance();           // 计算脉宽,并转换为距离
         for(a=100;a>0;a--)
         {   display(distance);
         }
         Init_Parameter();             // 参数重新初始化
    }
}
/***********************************************************************/
/* 函数名称:timer0                                                      */
/* 函数描述:T0 中断处理函数                                              */
/* 输入参数:无                                                          */
/* 参数描述:无                                                          */
/* 返回值:无                                                            */
/***********************************************************************/
void timer0(void) interrupt 1
{
  TF0 = 0;
  TL0 = 0x66;
  TH0 = 0xfc;
  count++;
  if(count == 18)// 超声波回声脉宽≤18ms
  {
    TR0 =0;
    TL0 = 0x66;
    TH0 = 0xfc;
    count = 0;
  }
}
/***********************************************************************/
```

超声波测距模块实物接线图如图 20-4 所示。实验时需要注意以下 2 点。

图 20-4　超声波测距模块实物接线图

(1)此模块不宜带电连接,若要带电连接,则再将模块的 GND 端口先连接,否则会影响模块的正常工作。

（2）测距时，被测物体的面积 ≥ 0.5m^2 且平面尽量要求平整，否则会影响测量结果。

20.4 技术拓展

20.4.1 超声波测距（LED 显示器显示改 I/O 端口）

超声波测距 LED 显示器显示改 I/O 端口实验程序如下。

```c
/********************BST-M51 实验开发板例程 *********************
* 平台：BST-M51 + Keil U4 + STC89C51
* 名称：超声波测距实验
* 晶振：11.059 2MHz
****************************************************************/
/************************ 包含头文件 ****************************/
#include <reg52.h>
/************************** 位定义 ******************************/
sbit echo = P1^0;                    // 回声接收端口
sbit trig = P1^1;                    // 超声触发端口

sbit wei1=P2^4;
sbit wei2=P2^5;
sbit wei3=P2^6;
sbit wei4=P2^7;
/************************** 宏定义 ******************************/
#define dula P0                      // 段选信号的锁存器控制
#define uchar unsigned char
#define uint unsigned int
/********************** 定义变量和数组 **************************/
long int distance=0;                 // 距离变量
uchar count;
unsigned char code table[]={0x3f,0x06,0x5b,0x4f,0x66,0x6d,0x7d,
                0x07,0x7f,0x6f,0x77,0x7c,0x39,0x5e,0x79,0x71};
unsigned char code table1[]={0xbf,0x86,0xdb,0xcf,0xe6,0xed,0xfd,
0x87,0xff,0xef,0x71};
/****************************************************************/
/* 函数名称：delay                                              */
/* 函数描述：延时函数                                           */
/* 输入参数：延时时间（ms）                                     */
/* 返回值：无                                                   */
/****************************************************************/
void delay (unsigned int xms)
{
   unsigned int i,j;
   for (i=xms;i>0;i--)               //i=xms 即延时 xms
       for (j=112;j>0;j--);
}
/****************************************************************/
/* 函数名称：display                                            */
/* 函数描述：LED 显示器显示函数                                 */
/* 输入参数：测试距离值的十倍                                   */
/* 返回值：无                                                   */
/****************************************************************/
void display (long int num)          // 显示程序
{
```

```c
uchar qian,bai,shi,ge;
if(num>10000)
{
   qian=15;//距离值上为百位
   bai=15;
   shi=10;
   ge=15;
}
else
{
   qian=num/1000;//距离值上为百位
   bai=(num/100)%10;//距离值上为十位
   shi=(num/10)%10;//距离值上为个位,并且显示时需要加小数点
   ge=num%10;//距离值上为小数位,并已经四舍五入
}
wei1=1;              // 显示千位
wei2=0;
wei3=0;
wei4=0;
P0=table[qian];
delay(2);

wei1=0;              // 显示百位
wei2=1;
wei3=0;
wei4=0;
P0=table[bai];
delay(2);

wei1=0;              // 显示十位
wei2=0;
wei3=1;
wei4=0;
P0=table1[shi];
delay(2);

wei1=0;              // 显示个位
wei2=0;
wei3=0;
wei4=1;
P0=table[ge];
delay(2);

}
```

20.4.2 超声波测距 LCD1602 显示

超声波测距 LCD1602 显示实验程序如下。

```c
/********************BST-M51 实验开发板例程 ***********************
*  平台: BST-M51 + Keil U4 + STC89C51
*  名称: 超声波测距实验(LCD1602 显示)
*  晶振: 1.0592MHz
***************************************************************/
/*************************** 包含头文件 ***************************/
#include <reg52.h>
#include "1602.h"
```

```c
/************************** 宏定义 **************************/
#define VELOCITY_30C 3495 //30℃时的声速,声速V= 331.5 + 0.6×温度;
#define VELOCITY_23C 3453 //23℃时的声速,声速V= 331.5 + 0.6×温度;
/************************** 位定义 **************************/
sbit INPUT  = P2^0;                   // 回声接收端口
sbit OUTPUT = P2^1;                   // 超声触发端口
sbit Beep   = P2^3 ;                  // 蜂鸣器
/*********************** 定义变量和数组 ***********************/
long int distance=0;                  // 距离变量
uchar table[]= "    Welcome to    ";
uchar table0[]= "   learn  BST-M51  ";
uchar table1[]= "There's no echo.";
uchar table2[]= "      BST-M51      ";
uchar table3[]= "Distance:";
uchar count;
/************************** 函数声明 **************************/
extern void initLCD();
extern void write_date (uchar date);
extern void write_com (uchar com);
extern void delay (uint x);
```

此项目附操作视频及代码资料。

Chapter 21

项目21 步进电机控制

项目目标

- 步进电机(电动机)是将电脉冲信号转变为角位移或线位移的开环控制元步进电机件。在非超载的情况下,电机的转速、停止的位置只取决于脉冲信号的频率和脉冲数,而不受负载变化的影响。当步进驱动器接收到一个脉冲信号,就驱动步进电机按设定的方向转动一个固定的角度,称为"步距角",它的旋转是以固定的角度一步一步运行的。可以通过控制脉冲个数来控制角位移量,从而达到准确定位的目的;同时可以通过控制脉冲频率来控制电机转动的速度和加速度,从而达到调速的目的。

建议学时

- 4学时。

知识要点

- 步进电机的工作原理。
- 结合按钮实现电机正反转。

技能掌握

- 了解步进电机的工作原理以及内部结构,掌握步进电机的驱动电路。

21.1 项目分析

步进电机是一种感应电机,它的工作原理是利用电子电路,将直流电变成分时供电的多相时序控制电流。用这种电流为步进电机供电,步进电机才能正常工作。驱动器就是为步进电机分时供电的多相时序控制器。

21.2 技术准备

21.2.1 步进电机简介

步进电机是一种将电脉冲转化为角位移的执行机构。当步进驱动器接收到一个脉冲信号,它就驱动步进电机按设定的方向转动一个步距角。通过控制脉冲来控制角位移量,可实现对电机的准确定位;通过控制脉冲频率来控制电机转动的速度和加速度,可实现对电机的调速。图 21-1 所示为 28BYJ-48 减速步进电机及驱动板实物图。

图 21-1　28BYJ-48 步进电机及驱动板实物图

21.2.2 步进电机转动原理及内部结构

减速步进电机 28BYJ-48 的内部结构原理图如图 21-2 所示,拆解图如图 21-3 所示。中间部分是转子,由一个永磁体组成,边上的是定子绕组。当定子的一个绕组通电时,将产生一个方向的电磁场,如果这个磁场的方向和转子磁场方向不在同一条直线上,那么定子和转子的磁场将产生一个扭力将定子扭转。

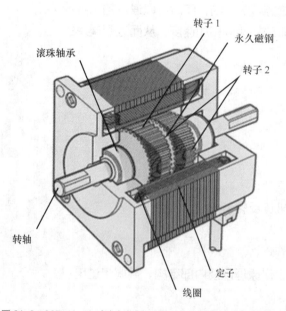

图 21-2　28BYJ-48 减速步进电机内部结构(与转轴平行方向的断面图)

(a)   (b)

图 21-3　28BYJ-48 减速步进电机拆解图

依次改变绕组的磁场，就可以使步进电机正转或反转。而改变磁场切换的时间间隔，就可以控制步进电机的速度了，这就是步进电机的驱动原理。

不管是两相四线，四相五线，四相六线步进电机，内部构造都是如此。

下面介绍几个步进电机相关概念。

（1）相数：产生不同对极 N，S 磁场的激磁线圈对数，常用 m 表示。

（2）拍数：完成 1 个磁场周期性变化所需脉冲数或导电状态，用 n 表示，或指电机转过 1 个齿距角所需脉冲数，以四相电机为例，有四相四拍运行方式即 AB-BC-CD-DA-AB，四相八拍运行方式即 A-AB-B-BC-C-CD-D-DA-A。

（3）步距角：对应 1 个脉冲信号，电机转子转过的角位移用 θ 表示。$\theta=360°$，以 28BYJ-48 四相五线步进电机为例，步距角为 5.625°。

21.2.3　ULN2003

由于步进电机的驱动电流较大，单片机不能直接驱动，一般都是使用 ULN2003 达林顿阵列驱动。ULN2003 内部是一个"非门"电路，包含 7 个单元，单独的每个单元驱动电流最大可达 500mA，9 脚可以悬空。ULN2003 与开发板及步进电机连接的原理图如图 21-4 所示。

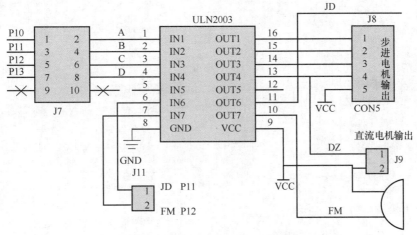

图 21-4　步进（直流）电机驱动模块电路

21.3　项目实施

21.3.1　单-双八拍

步进电机驱动单-双八拍工作方式实验程序如下。

```c
/**************************************************************
* 标题：步进电机试验一
* 通过本例程了解步进电机使用及驱动程序编写
单-双八拍工作方式：A-AB-B-BC-C-CD-D-DA
**************************************************************/
//接线一定要看步进电机杜邦线接法，步进电机模块上的"-"接开发板上的 GND 端口；步进电机模块上的"+"接开发板上的 VCC 端口
//IN1～IN4 端口分别接 P1.0～P1.3 端口
  #include "reg52.h"
  //Motor
sbit A = P1^0;           // 定义引脚
sbit b = P1^1;
sbit C = P1^2;
sbit D = P1^3;


/////////////////////////////////////////
//步进电机驱动
unsigned char MotorStep=0;   //步进电机步序
unsigned int  Speed=1,TIM,CT;

#define speed 12    // 调整速度数值不要设得太低，低了会引起震动

/*********************************************
步进电机初始化
*********************************************/
void InitMotor()
{
 A = 1;
 b = 1;
 C = 1;
 D = 1;
}
void SetMotor()
{
    switch(MotorStep)
    {
      case 0:
         if(TIM)  // A
          {
            A = 0;                //0xf1
            b = 1;
            C = 1;
            D = 1;
            MotorStep = 1;
           TIM=0;
          }
        break;
      case 1:                    // AB
         if(TIM)
          {
            A = 0;                //0xf3
            b = 0;
            C = 1;
            D = 1;
            MotorStep = 2;
          TIM=0;
```

```c
            }
        break;
        case 2:                 //B
            if(TIM)
            {
                A = 1;
                b = 0;          //0xf2
                C = 1;
                D = 1;
                MotorStep = 3;
            TIM=0;
            }
        break;
        case 3:                 //BC
            if(TIM)
            {
                A = 1;
                b = 0;          //0xf6
                C = 0;
                D = 1;
                MotorStep = 4;
            TIM=0;
            }
        break;
        case 4:                 //C
            if(TIM)
            {
                A = 1;
                b = 1;          //0xf4
                C = 0;
                D = 1;
                MotorStep = 5;
            TIM=0;
            }
        break;
    case 5:                     //CD
        if(TIM)
        {
            A = 1;
            b = 1;              //0xfc
            C = 0;
            D = 0;
            MotorStep = 6;
        TIM=0;
        }
        break;
    case 6:                     //D
        if(TIM)
        {
            A = 1;
            b = 1;              //0xf8
            C = 1;
            D = 0;
            MotorStep = 7;
```

```c
                TIM=0;
            }
        break;
    case 7:                        //DA
            if(TIM)
            {
                A = 0;
                b = 1;             //0xf9
                C = 1;
                D = 0;
                MotorStep = 0;
                TIM=0;
            }
        break;

    }
}
void system_Ini()
{
    TMOD|= 0x11;
    TH0=0xDC; //11.059 2MHz,定时 10ms
    TL0=0x00;
    IE = 0x8A;
    TR0 = 1;
}

main()
{
    system_Ini();
    InitMotor();
    while(1)
    {
        SetMotor();
    }
}

/*******************************************
定时中断延时
********************************************/
    void Tzd(void) interrupt 1
{
    TH0 = 0xDC;     //11.059 2MHz,定时 10ms
    TL0 = 0x00;

    TIM=1;
    }
```

21.3.2 加、减速

步进电机加、减速运行实验程序如下。

```
/*******************************************************************
*  标题:   步进电机试验三（加、减速运行）
*       通过本例程了解步进电机使用及驱动程序编写
单 - 双八拍工作方式：A-AB-B-BC-C-CD-D-DA
********************************************************************/
// 接线一定要看步进电机杜邦线接法,步进电机模块上的"-"接开发板上的 GND 端口;步进电机模块上的"+"接开发板上的
```

VCC 端口
```
//IN1～IN4 端口分别接 P1.0～P1.3 端口

  #include "reg52.h"

  void delay();

  //Motor
sbit F1 = P1^0;
sbit F2 = P1^1;
sbit F3 = P1^2;
sbit F4 = P1^3;

unsigned char code FFW[8]={0xfe,0xfc,0xfd,0xf9,0xfb,0xf3,0xf7,0xf6}; // 反转
unsigned char code FFZ[8]={0xf6,0xf7,0xf3,0xfb,0xf9,0xfd,0xfc,0xfe}; // 正转
unsigned int   K, rate;
/*******************************************************
步进电机驱动                                            *
*******************************************************/
void  motor_ffw()
  {
   unsigned char i;

      for (i=0; i<8; i++)
         {
          P1 = FFW[i]&0x1f;   // 取数据

           delay();             // 调节转速
         }
     }


/******************************************
   延时程序
******************************************/
  void delay()
{
   unsigned int k,t;
    t=rate;
   while(t--)
    {
     for(k=0; k<200; k++)
{ }
}
}
/******************************************
步进电机运行
******************************************/
void  motor_turn()
{
    unsigned char x;
    rate=0x09;
    x=0x40;
    do
     {
     motor_ffw();            // 加速
      rate--;
```

```c
    }while (rate!=0x01);
    do
     {
    motor_ffw();                    // 匀速
      x--;
    }while (x!=0x01);

    do
     {
     motor_ffw();                   // 减速
      rate++;
    }while (rate!=0x09);
}

main()
{
   while (1)
  {
  motor_turn();
   }
}
```

21.3.3 双四拍

步进电机驱动双四拍工作方式实验程序如下。

```
/*************************************************************************
*   标题：   步进电机试验四
*       通过本例程了解步进电机使用及驱动程序编写
*       双四拍工作方式：AB-BC-CD-DA
**************************************************************************/
// 接线一定要看步进电机杜邦线接法，步进电机模块上的"-"接开发板上的 GND 端口；步进电机模块上的"+"接开发板上的 VCC 端口
//IN1～IN4 端口分别接 P1.0～P1.3 端口


#include "reg52.h"
  //Motor
sbit F1 = P1^0;
sbit F2 = P1^1;
sbit F3 = P1^2;
sbit F4 = P1^3;
  ///////////////////////////////////////
// 步进电机驱动
unsigned char MotorStep=0;
unsigned int MotorTimer = 0;
unsigned int TIM,CT;

void InitMotor()
{
  F1 = 1;
  F2 = 1;
  F3 = 1;
  F4 = 1;
}
```

```c
void SetMotor()
{
//  if (Speed == 0) return;
    switch (MotorStep)
    {
       case 0:
            if (TIM)
            {
              F1 = 0;
              F2 = 0;
            F3 = 1;
            F4 = 1;
              MotorStep = 1;
              TIM=0;
              }
       break;
       case 1:
          if (TIM)
            {
              F1 = 1;
              F2 = 0;
              F3 = 0;
              F4 = 1;
              MotorStep = 2;
            TIM=0;
            }
         break;
         case 2:
           if (TIM)
             {
               F1 = 1;
               F2 = 1;
               F3 = 0;
               F4 = 0;
               MotorStep = 3;
             TIM=0;
             }
         break;
         case 3:
            if (TIM)
             {
               F1 = 0;
               F2 = 1;
               F3 = 1;
               F4 = 0;
               MotorStep = 0;
              TIM=0;
              }
         break;
    }


}
void system_Ini()
{
    TMOD|= 0x11;
```

```c
        TH0=0xDC;  //11.059 2MHz
        TL0=0x00;
    IE = 0x8A;
TR0  = 1;
}

main()
{
    system_Ini();
    InitMotor();
    while(1)
    {
        SetMotor();
    }
}

/*******************************************
*   定时中断延时
*******************************************/
    void Tzd(void) interrupt 1
{
    TH0 = 0xDC;        //11.059 2MHz,定时 10ms
    TL0 = 0x00;

    TIM=1;
    CT=0;
}
```

21.4 技术拓展

21.4.1 正、反转

步进电机正、反转实验程序如下。

```c
/***********************************************************************
*   标题：步进电机试验五（正转一下，反转一下）
*   通过本例程了解步进电机使用及驱动程序编写
*;     单-双八拍工作方式：A-AB-B-BC-C-CD-D-DA
***********************************************************************/
// 接线一定要看步进电机杜邦线接法，步进电机模块上的"-"接开发板上的 GND 端口；步进电机模块上的"+"接开发板上的 VCC 端口
//IN1～IN4 端口分别接 P1.0～P1.3 端口

    #include "reg52.h"
    void delay(unsigned int t);
    //Motor
sbit F1 = P1^0;
sbit F2 = P1^1;
sbit F3 = P1^2;
sbit F4 = P1^3;

unsigned char code FFW[8]={0xfe,0xfc,0xfd,0xf9,0xfb,0xf3,0xf7,0xf6}; // 反转
unsigned char code FFZ[8]={0xf6,0xf7,0xf3,0xfb,0xf9,0xfd,0xfc,0xfe}; // 正转
unsigned int  K;
```

```c
/************************************************************
步进电机驱动
*************************************************************/
void  motor_ffw()
 {
  unsigned char i;
  unsigned int  j;
  for (j=0; j<64; j++)
   {

     for (i=0; i<8; i++)
       {
         if (K==1) P1 = FFW[i]&0x1f;          // 取数据
         if (K==2) P1 = FFZ[i]&0x1f;
          delay (8);                          // 调节转速
       }
    }
  }


/***************************************
*   延时程序
****************************************/
  void delay(unsigned int t)
{
  unsigned int k;
  while (t--)
   {
     for (k=0; k<80; k++)
       { }
    }
 }



main()
  {
    while (1)
    {
    K=1;
    motor_ffw();
    K=2;
   motor_ffw();

   }
 }
```

21.4.2 速度调节

步进电机调速实验程序如下。

```
/*******************BST-V51 实验开发板例程 ***********************
*  平台：BST-M51 + Keil U4 + STC89C51RD
*  名称：步进电机调速
*  晶振：11.059 2MHz
*  实验效果：本程序用于测试 4 相步进电机常规驱动，4 个独立按键分别控制加速、减速、开启、停止
       LED 显示器显示 01～18 速度等级，数字越大，速度越快
****************************************************************/
```

```c
#include <reg52.h>

#define dula P0                    // 段选信号的锁存器控制
sbit wei1=P2^4;
sbit wei2=P2^5;
sbit wei3=P2^6;
sbit wei4=P2^7;

unsigned char code table[]={0x3f,0x06,0x5b,0x4f,0x66,0x6d,0x7d,
                            0x07,0x7f,0x6f,0x77,0x7c,0x39,0x5e,0x79,0x71};

sbit A1=P1^0;  // 定义步进电机连接端口
sbit B1=P1^1;
sbit C1=P1^2;
sbit D1=P1^3;
sbit k1 = P3^4;
sbit k2 = P3^5;
sbit k3 = P3^6;
sbit k4 = P3^7;
#define Coil_CD1 {A1=1;B1=1;C1=0;D1=0;}//CD 相通电，其他相断电
#define Coil_AD1 {A1=0;B1=1;C1=1;D1=0;}//AD 相通电，其他相断电
#define Coil_AB1 {A1=0;B1=0;C1=1;D1=1;}//AB 相通电，其他相断电
#define Coil_BC1 {A1=1;B1=0;C1=0;D1=1;}//BC 相通电，其他相断电
#define Coil_A1  {A1=0;B1=1;C1=1;D1=1;}//A 相通电，其他相断电
#define Coil_B1  {A1=1;B1=0;C1=1;D1=1;}//B 相通电，其他相断电
#define Coil_C1  {A1=1;B1=1;C1=0;D1=1;}//C 相通电，其他相断电
#define Coil_D1  {A1=1;B1=1;C1=1;D1=0;}//D 相通电，其他相断电
#define Coil_OFF {A1=1;B1=1;C1=1;D1=1;}// 全部断电

unsigned char Speed=1;
bit StopFlag;
void Display(unsigned char num);
void Init_Timer0(void);
/*-----------------------------------------------
延时函数，ms 级别
-----------------------------------------------*/
void delay(unsigned int xms)
{
  unsigned int i,j;
  for(i=xms;i>0;i--)          //i=xms 即延时 xms
     for(j=112;j>0;j--);
}
/*-----------------------------------------------
主函数
-----------------------------------------------*/
main()
{
  unsigned int i=512;// 旋转 1 周的时间
  //unsigned char num;
  Init_Timer0();
  Coil_OFF
  while(1) // 正向
  {
    if(k1 == 0) // 第 1 个按键，速度等级增加
    {
      delay(10);// 延时 10ms 消抖
```

```c
        if(k1 == 0)//再次判断按键是否被按下
        {
            while(k1 == 0);//直到判断按键松开
            if(Speed<18)
                    Speed++;
        }
    }
    else if(k2 == 0)//第2个按键,速度等级减小
    {
        delay(10);//延时10ms 消抖
        if(k2 == 0)//再次判断按键是否被按下
        {
            while(k2 == 0);//直到判断按键松开
            if(Speed>1)
                    Speed--;
        }
    }
    else if(k3 == 0)//第3个按键,暂停
    {
        Coil_OFF
        StopFlag=1;
    }
    else if(k4 == 0)//第4个按键,开启
        StopFlag=0;
    Display(Speed);
    }

}


/*-------------------------------------------------
LED 显示器显示函数,用于动态扫描 LED 显示器
-------------------------------------------------*/
void Display(unsigned char num)
{
    unsigned char shi,ge;

    shi = num/10;
    ge = num%10;

    wei1=0;                        //显示十位
    wei2=0;
    wei3=1;
    wei4=0;
    P0=table[shi];
    delay(2);

    wei1=0;                        //显示个位
    wei2=0;
    wei3=0;
    wei4=1;
    P0=table[ge];
    delay(2);
}
/*-------------------------------------------------
定时器初始化子程序
-------------------------------------------------*/
```

```c
void Init_Timer0(void)
{
  TMOD |= 0x01;        // 使用模式1，16位定时器，使用"|"符号可以在使用多个定时器时不受影响
//TH0=0x00;            // 给定初值
  //TL0=0x00;
  EA=1;                // 总中断打开
  ET0=1;               // 定时器中断打开
  TR0=1;               // 定时器开关打开
}
/*------------------------------------------------
定时器中断子程序
------------------------------------------------*/
void Timer0_isr(void) interrupt 1
{
  static unsigned char times,i;
  TH0=(65536-1000)/256;                 // 重新赋值 1ms
  TL0=(65536-1000)%256;

  if(!StopFlag)
   {
   if(times==(20-Speed))// 最大值18，所以最小间隔值为20-18=2
   {
    times=0;
    switch(i)
      {
       case 0:Coil_A1;i++;break;
       case 1:Coil_B1;i++;break;
       case 2:Coil_C1;i++;break;
       case 3:Coil_D1;i++;break;
       case 4:i=0;break;
       default:break;
      }
   }
  times++;
   }
}
```

21.4.3 自制秒表

利用步进电机自制秒表实验程序如下。

```c
/********************BST-V51实验开发板例程 ***********************
*  平台：BST-M51 + Keil U4 + STC89C5RD
*  名称：自制秒表
*  晶振：11.059 2MHz
*  实验效果：利用步进电机精准的转角制作圆形秒表
*****************************************************/
#include <reg52.h>

#define dula P0              // 段选信号的锁存器控制
sbit wei1=P2^4;
sbit wei2=P2^5;
sbit wei3=P2^6;
sbit wei4=P2^7;
sbit fm = P2^3;

unsigned char code table[]={0x3f,0x06,0x5b,0x4f,0x66,0x6d,0x7d,
                            0x07,0x7f,0x6f,0x77,0x7c,0x39,0x5e,0x79,0x71};
```

```c
sbit A1=P1^0;   //定义步进电机连接端口
sbit B1=P1^1;
sbit C1=P1^2;
sbit D1=P1^3;

#define Coil_AB1 {A1=0;B1=0;C1=1;D1=1;}//AB 相通电, 其他相断电
#define Coil_BC1 {A1=1;B1=0;C1=0;D1=1;}//BC 相通电, 其他相断电
#define Coil_CD1 {A1=1;B1=1;C1=0;D1=0;}//CD 相通电, 其他相断电
#define Coil_DA1 {A1=0;B1=1;C1=1;D1=0;}//D 相通电, 其他相断电
#define Coil_A1 {A1=0;B1=1;C1=1;D1=1;}//A 相通电, 其他相断电
#define Coil_B1 {A1=1;B1=0;C1=1;D1=1;}//B 相通电, 其他相断电
#define Coil_C1 {A1=1;B1=1;C1=0;D1=1;}//C 相通电, 其他相断电
#define Coil_D1 {A1=1;B1=1;C1=1;D1=0;}//D 相通电, 其他相断电
#define Coil_OFF {A1=1;B1=1;C1=1;D1=1;}// 全部断电

unsigned int count=0;
bit StopFlag;
void display (unsigned char num);
void Init_Timer0 (void);
/*------------------------------------------------
延时函数, ms 级别
------------------------------------------------*/
void delay(unsigned int xms)
{
   unsigned int i,j;
   for (i=xms;i>0;i--)              //i=xms 即延时 xms
       for (j=112;j>0;j--);
}
/*------------------------------------------------
主函数
------------------------------------------------*/
main()
{
  unsigned int time=0;
  Init_Timer0();
  fm = 1;
  Coil_OFF
  while (1)
   {
    if (count == 68)
    {
      count = 0;
      time++;
      if (time == 60)
          time = 0;
      fm = 0;
      delay(10);
      fm =1;
    }
    display(time);
   }
}


/*------------------------------------------------
LED 显示器显示函数, 用于动态扫描 LED 显示器
------------------------------------------------*/
```

```c
void display(unsigned char num)
{
    unsigned char shi,ge;
    
    shi = num/10;
    ge = num%10;
    
    wei1=0;                    // 显示十位
    wei2=0;
    wei3=1;
    wei4=0;
    P0=table[shi];
    delay(2);
    
    wei1=0;                    // 显示个位
    wei2=0;
    wei3=0;
    wei4=1;
    P0=table[ge];
    delay(2);
}
/*----------------------------------------------
定时器初始化子程序
----------------------------------------------*/
void Init_Timer0(void)
{
    TMOD |= 0x01;        // 使用模式1, 16位定时器, 使用"|"符号可以在使用多个定时器时不受影响
    TH0=0xcb;            // 重新赋值 (1/68)s
    TL0=0x0f;
    EA=1;                // 总中断打开
    ET0=1;               // 定时器中断打开
    TR0=1;               // 定时器开关打开
}
/*----------------------------------------------
定时器中断子程序
----------------------------------------------*/
void Timer0_isr(void) interrupt 1
{
    static unsigned char i;
    TH0=0xcb;            // 重新赋值 (1/68)s
    TL0=0x0f;
    
    switch(i)
    {
        case 0:Coil_A1;i++;break;
        case 1:Coil_AB1;i++;break;
        case 2:Coil_B1;i++;break;
        case 3:Coil_BC1;i++;break;
        case 4:Coil_C1;i++;break;
        case 5:Coil_CD1;i++;break;
        case 6:Coil_D1;i++;break;
        case 7:Coil_DA1;i=0;break;
        default:break;
    }
    count++;
}
```

此项目附操作视频及代码资料。

项目22
A/D-D/A模块

项目目标

- A/D-D/A模块口，A/D转换的功能是把模拟量电压转换为数字量电压。D/A转换的功能正好相反，就是将数字量转换为模拟量。

建议学时

- 4学时。

知识要点

- 芯片PCF8591。
- I²C总线原理。
- A/D-D/A模块程序烧录。

技能掌握

- 了解芯片PCF8591，掌握A/D-D/A模块程序烧录。

22.1 项目分析

本项目使用的 A/D-D/A 模块 PCF8591 是一个单片集成、单独供电、低功耗、8-bit CMOS 数据获取器件。PCF8591 具有 4 个模拟输入、1 个模拟输出和 1 个串行 I^{2}C 总线端口。PCF8591 的 3 个地址引脚 A0，A1 和 A2 可用于硬件地址编程，允许在同一个 I^{2}C 总线上接入 8 个 PCF8591，而无需额外的硬件。在 PCF8591 上，输入和输出地址、控制和数据信号都通过双线、双向 I^{2}C 总线以串行方式进行传输。

22.2 技术准备

22.2.1 PCF8591 介绍

PCF8591 是具有 I^{2}C 总线接口的 8 位 A/D 及 D/A 转换器其原理图如图 22-1 所示。有 4 路 A/D 转换输入，1 路 D/A 模拟输出。这就是说，它既可以作 A/D 转换也可以作 D/A 转换。A/D 转换为逐次比较型。电源电压典型值为 5V。

AIN0 ~ AIN3：模拟信号输入端。

A0 ~ A3：引脚地址端。

VSS：电源负极。

SDA，SCL：I^{2}C 总线的数据线、时钟线。

OSC：外部时钟输入端，内部时钟输出端。

EXT：内部、外部时钟选择线，使用内部时钟时，EXT 接地。

AGND：模拟信号地。

VREF：基准电源端。

AOUT：D/A 转换输出端。

VDD：电源端（2.5 ~ 6V）。

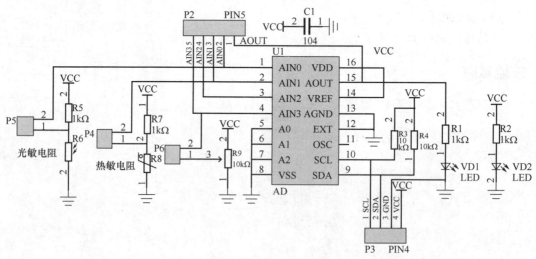

图 22-1 PCF8591 原理图

22.2.2 PCF8591 的器件地址与控制寄存器

PCF8591 采用典型的 I²C 总线接口器件寻址方法,即总线地址由器件地址、引脚地址和方向位组成,见表 22-1。规定 A/D 器件地址为 1001。引脚地址 A2A1A0,其值由用户选择,因此 I²C 系统中最多可接 2^3=8 个具有 I²C 总线接口的 A/D 器件。地址的最后一位为方向位 R/W,当主控器对 A/D 器件进行读操作时为 1,进行写操作时为 0。总线操作时,由器件地址、引脚地址和方向位组成的从地址为主控器发送第 1 字节。PCF8591 控制寄存器见表 22-2,器件地址见表 22-3。

表 22-1 PCF8591 总线地址

MSB							LSB
D7	D6	D5	D4	D3	D2	D1	D0

D7 ~ D4 规定为 1001。
D3 ~ D1 分别是 A2,A1,A0 的电平,原理图上面是全部接地,所以为 000。
D0 为方向设置,当为 1 时进行读操作,当为 0 时进行写操作。

表 22-2 PCF8591 控制寄存器

MSB							LSB
0	×	×	×	0	×	×	×
D7	D6	D5	D4	D3	D2	D1	D0

表 22-3 PCF8591 的器件地址

D1,D0	A/D 通道选择 00 通道 0;01 通道 1;10 通道 2;11 通道 3
D2	自动增益选择(有效位为 1)
D5,D4	输入模式选择:00 为 4 路单数输入;01 为 3 路差分输入;10 为单端与差分配合输入;11 为模拟输入有效
D6	模拟输出使能位(使能为 1)

22.2.3 I²C 总线的数据传送

1. I²C 总线数据传送简介

I²C 总线数据传送的起始和终止信号都是由主机发出的,在起始信号产生后,总线就处于被占用的状态;在终止信号产生后,总线就处于空闲状态。

在起始信号后必须传送一个从机的地址(7 位),第 8 位是数据的传送方向位(R/T),用"0"表示主机发送数据(T),"1"表示主机接收数据(R)。每次数据传送总是由主机产生的终止信号结束。但是,若主机希望继续占用总线进行新的数据传送,则可以不产生终止信号,马上再次发出起始信号,对另一从机进行寻址。

主机可以采用不带 I²C 总线接口的单片机,如 80C51,AT89C2051 等单片机,利用软件实现 I²C 总线的数据传送,即软件与硬件结合的信号模拟。为了保证数据传送的可靠性,标准的 I²C 总线的数据传送有严格的时序要求。I²C 总线的起始信号、终止信号、发送"0"及发送"1"的模拟时序图如图 22-2 所示。

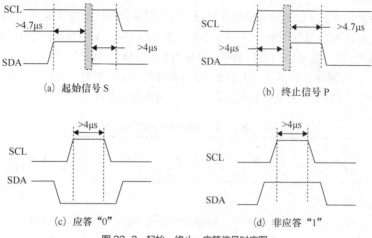

图 22-2　起始、终止、应答信号时序图

2. I²C 总线数据传送相关程序

（1）起始信号。

```
/*************************************************************
启动总线函数
函数原型: void  Start_I2c();
功能: 启动 I²C 总线，即发送 I²C 起始条件
**************************************************************/
void Start_I2c()
{
  sda=1;            /*发送起始条件的数据信号*/
  _nop_();
  scl=1;
  _nop_();          /*起始条件建立时间>4.7μs, 延时*/
  _nop_(); _nop_(); _nop_(); _nop_();
  sda=0;            /*发送起始信号*/
  _nop_();          /*起始条件锁定时间>4μs*/
  _nop_(); _nop_(); _nop_(); _nop_();
  scl=0;            /*开启 I²C 总线，准备发送或接收数据 */
  _nop_(); _nop_();
}
```

（2）终止信号。

```
/*************************************************************
结束总线函数
函数原型: void  Stop_I2c();
功能: 结束 I²C 总线，即发送 I²C 结束条件
**************************************************************/
void Stop_I2c()
{
  sda=0;          /*发送结束条件的数据信号*/
  _nop_();        /*发送结束条件的时钟信号*/
  scl=1;          /*结束条件建立时间>4μs*/
  _nop_(); _nop_(); _nop_(); _nop_(); _nop_();
  sda=1;          /*发送 I²C 总线结束信号*/
  _nop_(); _nop_(); _nop_(); _nop_();
}
```

（3）应答信号。

```
/***********************************************************
应答子函数
函数原型： void Ack_I2c（bit a）;
功能：主控器进行信号应答（可以是应答或非应答，由位参数 a 决定）
***********************************************************/
void Ack_I2c(bit a)
{

  if（a==0）sda=0;                /*在此发出应答或非应答信号 */
  else sda=1;
  _nop_(); _nop_(); _nop_();
  scl=1;
  _nop_(); /*时钟低电平周期大于 4μs*/
  _nop_(); _nop_(); _nop_(); _nop_();
  scl=0;                          /*清时钟线，开启 I²C 总线以便继续接收 */
  _nop_(); _nop_();
}
```

每一字节必须保证是 8 位长度。数据传送时，先传送最高位（MSB），每一个被传送的字节后面都必须跟随一位应答位（即 1 帧，共 9 位）。字节传送与应答时序图如图 22-3 所示。

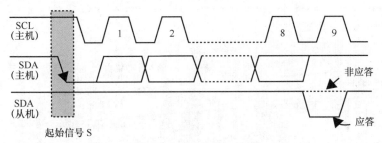

图 22-3　字节传送与应答时序图

由于某种原因从机不对主机寻址信号应答（如从机正在进行实时性的处理工作而无法接收总线上的数据）时，必须将其数据线置于高电平，而由主机产生一个终止信号，以结束总线的数据传送。

如果从机对主机进行了应答，但在数据传送一段时间后，无法继续接收更多的数据时，从机可以通过对无法接收的第 1 个数据字节的"非应答"通知主机，主机则应发出终止信号，以结束数据的继续传送。

当主机接收数据时，它收到最后一个数据字节后，必须向从机发出一个结束传送的信号。这个信号是由对从机的"非应答"来实现的。然后，从机释放 SDA 线，以允许主机产生终止信号。

（4）发送 1 字节。

```
/***********************************************************
字节数据发送函数
函数原型： void  I2C_SendByte（UCHAR c）;
功能：将数据 c 发送出去，可以是地址，也可以是数据，发送完毕后等待应答，并对此状态位进行操作，ack=1，发送数据正常；
ack=0，被控器件无应答或损坏（即不应答，或非应答，都使 ack=0）
***********************************************************/
void I2C_SendByte(unsigned char  c)
{
  unsigned char  i;
  for（i=0;i<8;i++) /*要传送的数据长度为 8 位 */
   {
    if（(c<<i)&0x80) sda=1;    /*判断发送位 */
      else  sda=0;
```

```c
        _nop_();
        scl=1;                    /*置时钟线为高电平,通知被控器件开始接收数据位*/
        _nop_();                  /*保证时钟高电平周期>4μs*/
        _nop_(); _nop_(); _nop_(); _nop_();
        scl=0;
    }
    _nop_(); _nop_();
    sda=1;                        /*8位发送完后释放数据线,准备接收应答位*/
    _nop_(); _nop_();
    scl=1;
    _nop_(); _nop_(); _nop_();
    if(sda==1) ack=0;
      else ack=1;                 /*判断是否接收到应答信号*/
    scl=0;
    _nop_(); _nop_();
}
```

(5) 接收 1 字节。

```c
/*****************************************************
字节数据接收函数
函数原型: UCHAR I2C_RcvByte();
功能:用来接收从器件传来的数据,并判断总线错误(不发应答信号),发完后请用应答函数应答从机
*****************************************************/
unsigned char I2C_RcvByte()
{
    unsigned char retc=0,i;
    sda=1;                        /*置数据线为输入方式*/
    for(i=0;i<8;i++)
    {
        _nop_();
        scl=0;                    /*置时钟线为低电平,准备接收数据位*/
        _nop_();                  /*时钟低电平周期>4.7μs*/
        _nop_(); _nop_(); _nop_(); _nop_();
        scl=1;                    /*置时钟线为高电平,使数据线上数据有效*/
        _nop_(); _nop_();
        retc=retc<<1;
        if(sda==1) retc=retc+1;   /*读数据位,接收的数据位放入retc中*/
        _nop_(); _nop_();
    }
    scl=0;
    _nop_(); _nop_();
    return(retc);
}
```

(6) PCF8591 的写入。

第 1 字节是器件地址和读写控制。

第 2 字节被存到控制寄存器,用于控制器件功能。

第 3 字节被存储到 DAC 数据寄存器,并使用芯片上 D/A 转换器转换成对应的模拟电压(不输入 D/A 时,不用输入)。

PCF8591 发送 1 字节程序如下

```c
/*****************************************************
* 函数名: Pcf8591_SendByte
* 函数功能: 写入一个控制命令
* 输入: addr(器件地址),channel(转换通道)
* 输出: 无
```

```
**************************************************/
bit PCF8591_SendByte (unsigned char addr,unsigned char channel)
{
    Start_I2c();                            // 启动总线
    I2C_SendByte (addr);                    // 发送器件地址
    if (ack==0) return (0);
    I2C_SendByte (0x40|channel);            // 发送控制字节
    if (ack==0) return (0);
    Stop_I2c();                             // 结束总线
    return (1);
}
```

(7) PCF8591 的读取。

读取的第 1 字节是包含上一次转换结果。将上 1 字节读取时，才开始进行本次转换的采样。读取的第 2 字节才是本次的转换结果。所以读取转换结果的步骤是，发送转换命令，将上次的结果读走，经短时等待，然后读取结果。

① PCF8591 读取 1 字节的程序。

```
/************************************************
* 函数名：PCF8591_RcvByte
* 函数功能：读取 1 个转换值
* 输入：
* 输出：dat
**************************************************/
unsigned char PCF8591_RcvByte (unsigned char addr)
{   unsigned char dat;
    Start_I2c();                            // 启动总线
    I2C_SendByte (addr+1);                  // 发送器件地址
    if (ack==0) return (0);
    dat=I2C_RcvByte();                      // 读取数据
    Ack_I2c (1);                            // 发送非应答信号
    Stop_I2c();                             // 结束总线
    return (dat);
}
```

② PCF8591 发送 1 次转换的程序。

```
/************************************************
* 函数名：Pcf8591_DaConversion
* 函数功能：PCF8591 输出端输出模拟量
* 输入：addr（器件地址），channel（转换通道），value（转换的数值）
* 输出：无
**************************************************/
bit Pcf8591_DaConversion (unsigned char addr,unsigned char channel, unsigned char Val)
{
    Start_I2c();                            // 启动总线
    I2C_SendByte (addr);                    // 发送器件地址
    if (ack==0) return (0);
    I2C_SendByte (0x40|channel);            // 发送控制字节
    if (ack==0) return (0);
    I2C_SendByte (Val);                     // 发送 DAC 的数值
    if (ack==0) return (0);
    Stop_I2c();                             // 结束总线
    return (1);
}
```

A/D-D/A 实验模块实物接线图如图 22-4 所示。注意，务必按照图示连接开发板。

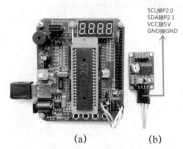

图 22-4　A/D-D/A 模块实验实物连接图

此项目附操作视频及代码资料。

22.3　项目实施

A/D 模块串口显示实验程序如下。

```
/*******************BST-V51 实验开发板例程 *********************
*   平台:BST-M51 + Keil U4 + STC89C51
*   名称:A/D-D/A 模块实验（串口显示）
*   晶振:11.0592MHz
***************************************************************/
#include<reg52.h>        // 包含单片机寄存器的头文件
#include <intrins.h>
#define   AddWr 0x90     //PCF8591 地址

// 变量定义
unsigned char AD_CHANNEL=0;
unsigned long xdata   LedOut[8];
unsigned char    D[32];

sbit scl=P2^0;           //I²C  时钟
sbit sda=P2^1;           //I²C  数据
bit ack;                 /*应答标志位*/
unsigned char date;

/****************************************************************
启动总线函数
函数原型:void  Start_I2c();
功能:       启动 I²C 总线,即发送 I²C 起始条件
****************************************************************/
void Start_I2c()
{
  sda=1;           /*发送起始条件的数据信号*/
  _nop_();
  scl=1;
  _nop_();         /*起始条件建立时间>4.7μs,延时*/
  _nop_();
  _nop_();
  _nop_();
  _nop_();
  sda=0;           /*发送起始信号*/
  _nop_();         /* 起始条件锁定时间>4μs*/
```

```
       _nop_();
       _nop_();
       _nop_();
       _nop_();
       scl=0;           /*钳住I²C总线,准备发送或接收数据 */
       _nop_();
       _nop_();
}

/******************************************************************
结束总线函数
函数原型: void  Stop_I2c();
功能:       结束I²C总线,即发送I²C结束条件
******************************************************************/
void Stop_I2c()
{
sda=0;         /*发送结束条件的数据信号*/
 _nop_();      /*发送结束条件的时钟信号*/
scl=1;         /*结束条件建立时间大于4μs*/
_nop_();
 _nop_();
 _nop_();
 _nop_();
 _nop_();
sda=1;         /*发送I²C总线结束信号*/
 _nop_();
 _nop_();
 _nop_();
 _nop_();
}

/******************************************************************
字节数据发送函数
函数原型: void  I2C_SendByte(UCHAR c);
功能:将数据c发送出去,可以是地址,也可以是数据,发送完毕后等待应答,并对此状态位进行操作,ack=1,发送数据
正常.; ack=0,被控器件无应答或损坏(即不应答,或非应答,都使ack=0)
******************************************************************/
void  I2C_SendByte(unsigned char  c)
{
  unsigned char  i;

  for(i=0;i<8;i++) /*要传送的数据长度为8位*/
    {
      if((c<<i)&0x80) sda=1;  /*判断发送位*/
       else  sda=0;
      _nop_();
      scl=1;                  /*置时钟线为高电平,通知被控器件开始接收数据位*/
       _nop_();
       _nop_();               /*保证时钟高电平周期>4μs*/
       _nop_();
       _nop_();
       _nop_();
      scl=0;
    }

     _nop_();
```

```c
   _nop_();
      sda=1;                      /*8位发送完后释放数据线，准备接收应答位*/
  _nop_();
     _nop_();
     scl=1;
     _nop_();
     _nop_();
     _nop_();
     if(sda==1)ack=0;
        else ack=1;               /*判断是否接收到应答信号*/
     scl=0;
     _nop_();
     _nop_();
}

/***************************************************************
字节数据接收函数
函数原型：UCHAR   I2C_RcvByte();
功能：用来接收从器件传来的数据，并判断总线错误（不发应答信号），发完后请用应答函数应答从机
****************************************************************/
unsigned char   I2C_RcvByte()
{
  unsigned char  retc=0,i;
  sda=1;                          /*置数据线为输入方式*/
  for(i=0;i<8;i++)
     {
       _nop_();
       scl=0;                     /*置时钟线为低电平，准备接收数据位*/
       _nop_();
       _nop_();                   /*时钟低电平周期>4.7μs*/
       _nop_();
       _nop_();
       _nop_();
       scl=1;                     /*置时钟线为高电平，使数据线上数据有效*/
       _nop_();
       _nop_();
       retc=retc<<1;
       if(sda==1)retc=retc+1;     /*读数据位，接收的数据位放入retc中*/
       _nop_();
       _nop_();
      }
  scl=0;
  _nop_();
  _nop_();
  return(retc);
}

/***************************************************************
应答子函数
函数原型： void Ack_I2c(bit a);
功能：主控器进行信号应答（可以是应答或非应答，由位参数a决定）
****************************************************************/
void Ack_I2c(bit a)
{
  if(a==0)sda=0;                  /*在此发出应答或非应答信号*/
    else sda=1;                   /*0为发出应答，1为非应答信号*/
```

```c
    _nop_();
    _nop_();
    _nop_();
    scl=1;
    _nop_();
    _nop_();                    /*时钟低电平周期>4μs*/
    _nop_();
    _nop_();
    _nop_();
    scl=0;                      /*清时钟线,钳住I²C总线以便继续接收*/
    _nop_();
    _nop_();
}

/***********************************************
* 函数名: Pcf8591_DaConversion
* 函数功能: PCF8591的输出端输出模拟量
* 输入: addr(器件地址)、channel(转换通道)、value(转换的数值)
* 输出      : 无
*********************** ***********************************/
bit Pcf8591_DaConversion(unsigned char addr,unsigned char channel, unsigned char Val)
{
    Start_I2c();                    //启动总线
    I2C_SendByte(addr);             //发送器件地址
    if(ack==0) return(0);
    I2C_SendByte(0x40|channel);     //发送控制字节
    if(ack==0) return(0);
    I2C_SendByte(Val);              //发送DAC的数值
    if(ack==0) return(0);
    Stop_I2c();                     //结束总线
    return(1);
}


/***********************************************
* 函数名: Pcf8591_SendByte
* 函数功能: 写入一个控制命令
* 输入: addr(器件地址)、channel(转换通道)
* 输出: 无
***********************************************/
bit PCF8591_SendByte(unsigned char addr,unsigned char channel)
{
    Start_I2c();                    //启动总线
    I2C_SendByte(addr);             //发送器件地址
    if(ack==0) return(0);
    I2C_SendByte(0x40|channel);     //发送控制字节
    if(ack==0) return(0);
    Stop_I2c();                     //结束总线
    return(1);
}


/***********************************************
* 函数名: PCF8591_RcvByte
* 函数功能: 读取1个转换值
* 输入:
* 输出: dat
***********************************************/
```

```c
unsigned char PCF8591_RcvByte(unsigned char addr)
{
    unsigned char dat;

    Start_I2c();                    // 启动总线
    I2C_SendByte(addr+1);           // 发送器件地址
    if(ack==0) return(0);
    dat=I2C_RcvByte();              // 读取数据0

    Ack_I2c(1);                     // 发送非应答信号
    Stop_I2c();                     // 结束总线
    return(dat);
}
/*-----------------------------------------------
串口初始化函数
-----------------------------------------------*/
void init_com(void)
{
  EA=1;                 // 开总中断
  ES=1;                 // 允许串口中断
  ET1=1;                // 允许定时器T1中断
  TMOD=0x20;            // 定时器T1，在方式2中断产生波特率
  PCON=0x00;            //SMOD=0
  SCON=0x50;            // 方式1 由定时器控制
  TH1=0xfd;             // 波特率设置为9 600bit/s
  TL1=0xfd;
  TR1=1;                // 开定时器T1，运行控制位

}
/*-----------------------------------------------
延时函数
-----------------------------------------------*/
void delay(unsigned char i)
{
  unsigned char j,k;
  for(j=i;j>0;j--)
    for(k=125;k>0;k--);
}
/*-----------------------------------------------
把读取值转换成一个一个的字符，送串口显示
-----------------------------------------------*/
void To_ascii(unsigned char num)
{
    SBUF=num/100+'0';
    delay(200);
    SBUF=num/10%10+'0';
    delay(200);
    SBUF=num%10+'0';
    delay(200);
}
/*-----------------------------------------------
主函数
-----------------------------------------------*/
main()
{
```

```c
init_com();
while(1)
{

/******** 以下A/D-D/A转换 *************/

    switch(AD_CHANNEL)
    {
        case 0: PCF8591_SendByte(AddWr,1);
            D[0]=PCF8591_RcvByte(AddWr);    //ADC0,模/数转换1,光敏电阻
            break;

        case 1: PCF8591_SendByte(AddWr,2);
            D[1]=PCF8591_RcvByte(AddWr);    //ADC1,模/数转换2,热敏电阻
            break;

        case 2: PCF8591_SendByte(AddWr,3);
            D[2]=PCF8591_RcvByte(AddWr);    //ADC2,模/数转换3,悬空
            break;

        case 3: PCF8591_SendByte(AddWr,0);
            D[3]=PCF8591_RcvByte(AddWr);    //ADC3,模/数转换4,可调0~5V
            break;

        case 4: Pcf8591_DaConversion(AddWr,0, D[4]);  //DAC    数/模转换
        break;

    }

D[4]=D[3];    // 把模拟输入采样的信号通过数/模转换输出

if(++AD_CHANNEL>4) AD_CHANNEL=0;

/******** 以下将A/D转换的值通过串口发送出去 *************/
delay(200);
 To_ascii(D[0]);
 SBUF=' ';
 delay(200);
 To_ascii(D[1]);
 SBUF=' ';
 delay(200);
 To_ascii(D[2]);
 SBUF=' ';
delay(200);
 To_ascii(D[3]);
 SBUF='\n';
 delay(200);
if(RI)
{
 date=SBUF;     // 单片机接收
 SBUF=date;     // 单片机发送
 RI=0;
}
 }
}
```

22.4 技术拓展

22.4.1 D/A 输出模块

D/A 输出模块实验程序如下。

```
/********************BST-V51 实验开发板例程 ************************
*   平台: BST-M51 + Keil U4 + STC89C51
*   名称: D/A 输出模块实验
*   晶振: 11.059 2MHz
*   实验效果: 调节模块上的可调电阻,调节 LED 的亮度
*****************************************************************/
#include<reg52.h>         // 包含单片机寄存器的头文件
#include <intrins.h>
#define  AddWr 0x90       //PCF8591 地址
// 变量定义
unsigned char AD_CHANNEL;
unsigned long xdata  LedOut[8];
unsigned char   D[32];

sbit scl=P2^0;            //I²C 时钟
sbit sda=P2^1;            //I²C 数据
bit ack;                  /*应答标志位*/


unsigned char date;


/***********************************************************
启动总线函数
函数原型: void  Start_I2c();
功能: 启动 I²C 总线,即发送 I²C 起始条件
************************************************************/
```

22.4.2 A/D 模块(LCD1602 显示)

A/D 模块 LCD1602 显示实验程序如下。

```
/********************BST-V51 实验开发板例程 ************************
*   平台: BST-M51 + Keil U4 + STC89C52RC
*   名称: A/D 模块实验
*   晶振: 11.059 2MHz
*   实验效果: 在 LCD 1602 上显示各通道的电压值
*****************************************************************/
#include<reg52.h>         // 包含单片机寄存器的头文件
#include <intrins.h>

#define  AddWr 0x90       //PCF8591 地址

// 变量定义
unsigned char AD_CHANNEL;
sbit scl=P2^0;            //I²C 时钟
sbit sda=P2^1;            //I²C 数据
bit ack;                  /*应答标志位*/

sbit RS = P1^0;//Pin4
sbit RW = P1^1; //Pin5
sbit E  = P2^5;//Pin6
```

```c
#define Data  P0        // 数据端口
unsigned char TempData[8];
unsigned char  SecondLine[]= "              ";
unsigned char  FirstLine[] = "              ";
/***************************************************************
启动总线函数
函数原型: void  Start_I2c();
功能: 启动 I²C 总线，即发送 I²C 起始条件
****************************************************************/
void Start_I2c()
{
  sda=1;              /*发送起始条件的数据信号*/
  _nop_();
  scl=1;
  _nop_();            /*起始条件建立时间>4.7μs,延时*/
  _nop_();
  _nop_();
  _nop_();
  _nop_();
  sda=0;              /*发送起始信号*/
  _nop_();            /* 起始条件锁定时间>4μs*/
  _nop_();
  _nop_();
  _nop_();
  _nop_();
  scl=0;              /*钳住 I²C 总线，准备发送或接收数据 */
  _nop_();
  _nop_();
}


/***************************************************************
结束总线函数
函数原型 : void  Stop_I2c();
功能:      结束 I²C 总线，即发送 I²C 结束条件
****************************************************************/
/*-----------------------------------------------
主函数
-----------------------------------------------*/
main()
{
   unsigned char ADtemp;                   // 定义中间变量
   InitLcd();
    mDelay(20);

  while(1)
  {

  /******** 以下 A/D-D/A 转换 ************/

     switch(AD_CHANNEL)
    {
       case 0: PCF8591_SendByte(AddWr,1);
            ADtemp = PCF8591_RcvByte(AddWr);   //ADC0,模/数转换1,光敏电阻
            TempData[0]=ADtemp/50;// 处理 0 通道电压显示
            TempData[1]=(ADtemp%50)/10;
            break;
```

```
        case 1: PCF8591_SendByte(AddWr,2);
                ADtemp=PCF8591_RcvByte(AddWr);   //ADC1,模/数转换2,热敏电阻
                TempData[2]=ADtemp/50;//处理1通道电压显示
                TempData[3]=(ADtemp%50)/10;
                break;

        case 2: PCF8591_SendByte(AddWr,3);
        ADtemp=PCF8591_RcvByte(AddWr);   //ADC2,模/数转换3,悬空
                TempData[4]=ADtemp/50;//处理2通道电压显示
                TempData[5]=(ADtemp%50)/10;
                break;

        case 3: PCF8591_SendByte(AddWr,0);
                ADtemp=PCF8591_RcvByte(AddWr);   //ADC3,模/数转换4,可调0～5V
                TempData[6]=ADtemp/50;//处理3通道电压显示
                TempData[7]=(ADtemp%50)/10;
                break;

    case 4: Pcf8591_DaConversion(AddWr,0,ADtemp); //DAC    数/模转换
            break;
    }
    if(++AD_CHANNEL>4) AD_CHANNEL=0;
  disp();
    }
}
```

此项目附操作视频及代码资料。

Chapter 23

项目23
火焰报警器

项目目标

- 火焰报警器中的主要部件火焰传感器（即红外接收二极管）是一种专门用来搜寻火源的传感器。火焰传感器也可以用来检测光线的亮度，只是它对火焰更敏感。火焰传感器使用特制的红外线接收管来检测火焰，然后把火焰的亮度换成电平信号传送给中央处理器，中央处理器据此做出反应。

建议学时

- 4学时。

知识要点

- 火焰传感器接线原理。
- 调节火焰传感器。

技能掌握

- 了解火焰传感器的工作原理，掌握对火焰传感器数据的检测方法。

23.1 项目分析

在熟悉项目 22 所学 A/D-D/A 模块的基础上，利用面包板接线，AIN0 接入光敏电阻，AIN1 接入热敏电阻，用串口观察其电压变化。

光敏电阻器是利用半导体的光电导效应制成的一种电阻值随入射光的强弱而改变的电阻器，又称为光电导探测器。有 2 种特性截然相反的光敏电阻器，一种是随入射光增强，电阻值减小；随入射光减弱，电阻值增大。另一种是随入射光增强，电阻值增大；随入射光减弱，电阻值减小。

热敏电阻器也是敏感元件的一类，亦属半导体器件。按照温度系数不同，分为正温度系数热敏电阻器（PTC）和负温度系数热敏电阻器（NTC）。热敏电阻器的典型特点是对温度敏感，不同的温度下呈现出不同的电阻值。正温度系数热敏电阻器（PTC）在温度越高时电阻值越大，负温度系数热敏电阻器（NTC）在温度越高时电阻值越低。

23.2 技术准备

23.2.1 光、热敏电阻拓展接线原理

光、热敏电阻接线原理图如图 23-1 所示。

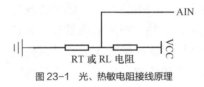

图 23-1 光、热敏电阻接线原理

23.2.2 火焰传感器介绍

火焰传感器实物如图 23-2 所示，利用红外线对火焰非常敏感的特点，使用特制的红外线接收管来检测火焰，然后把火焰的亮度转化为高低变化的电平信号，输入到中央处理器，中央处理器根据信号的变化做出相应的程序处理。一般可以用来探测火源或其他一些波长在 700 ~ 1 000nm 范围内的热源，在探测到此频段波长时，阻值变大，电路不导通。其他常态下，电路导通。

图 23-2 火焰传感器实物图

23.3 项目实施

A/D 模块串口读取实验程序如下。

```
/*********************BST-M51 实验开发板例程 *********************
*  平台：BST-M51 + Keil U4 + STC89C51
*  名称：A/D 串口读取实验
```

```c
*  晶振：11.059 2MHz
*****************************************************************/
#include<reg52.h>         // 包含单片机寄存器的头文件
#include <intrins.h>
#define  AddWr 0x90       //PCF8591 地址
// 变量定义
unsigned char AD_CHANNEL=0;
unsigned char  D[32];

sbit scl=P2^0;            //I²C 时钟
sbit sda=P2^1;            //I²C 数据
bit ack;                  /*应答标志位*/
unsigned char date;
/*****************************************************************
启动总线函数
函数原型：void  Start_I2c();
功能：    启动 I²C 总线，即发送 I²C 起始条件
*****************************************************************/
void Start_I2c()
{
sda=1;                    /*发送起始条件的数据信号*/
  _nop_();
scl=1;
  _nop_();                /*起始条件建立时间>4.7μs，延时*/
  _nop_();
  _nop_();
  _nop_();
  _nop_();
sda=0;                    /*发送起始信号*/
  _nop_();                /* 起始条件锁定时间>4μs*/
  _nop_();
  _nop_();
  _nop_();
  _nop_();
scl=0;                    /*钳住 I²C 总线，准备发送或接收数据 */
  _nop_();
  _nop_();
}


/*****************************************************************
结束总线函数
函数原型：void  Stop_I2c();
功能：结束 I²C 总线，即发送 I²C 结束条件
*****************************************************************/
void Stop_I2c()
{
sda=0;                    /*发送结束条件的数据信号*/
  _nop_();                /*发送结束条件的时钟信号*/
scl=1;                    /*结束条件建立时间>4μs*/
_nop_();
  _nop_();
  _nop_();
  _nop_();
  _nop_();
sda=1;                    /*发送 I²C 总线结束信号*/
  _nop_();
```

```c
    _nop_();
    _nop_();
    _nop_();
}

/*************************************************************
字节数据发送函数
函数原型：void  I2C_SendByte(UCHAR c);
功能：将数据c发送出去，可以是地址，也可以是数据，发送完毕后等待应答，并对此状态位进行操作 ack=1,发送数据正常；
ack=0，被控器件无应答或损坏（即不应答，或非应答，都使ack=0）
*************************************************************/
void  I2C_SendByte(unsigned char  c)
{
  unsigned char  i;
  for(i=0;i<8;i++)    /*要传送的数据长度为8位*/
    {
     if((c<<i)&0x80) sda=1;    /*判断发送位*/
      else  sda=0;
     _nop_();
     scl=1;                    /*置时钟线为高电平，通知被控器件开始接收数据位*/
     _nop_();
     _nop_();                  /*保证时钟高电平周期>4μs*/
     _nop_();
     _nop_();
     _nop_();
     scl=0;
    }
    _nop_();
   _nop_();
     sda=1;                    /*8位发送完后释放数据线，准备接收应答位*/
   _nop_();
     _nop_();
     scl=1;
     _nop_();
     _nop_();
     _nop_();
     if(sda==1)ack=0;
       else ack=1;             /*判断是否接收到应答信号*/
     scl=0;
     _nop_();
     _nop_();
 }

/*************************************************************
字节数据接收函数
函数原型：UCHAR  I2C_RcvByte();
功能：用来接收从器件传来的数据，并判断总线错误（不发应答信号），发完后请用应答函数应答从机
*************************************************************/
unsigned char  I2C_RcvByte()
{
  unsigned char  retc=0,i;
  sda=1;                       /*置数据线为输入方式*/
  for(i=0;i<8;i++)
    {
      _nop_();
```

```c
            scl=0;                          /*置时钟线为低电平,准备接收数据位*/
            _nop_();
            _nop_();                        /*时钟低电平周期>4.7μs*/
            _nop_();
            _nop_();
            _nop_();
            scl=1;                          /*置时钟线为高电平,使数据线上数据有效*/
            _nop_();
            _nop_();
            retc=retc<<1;
            if(sda==1) retc=retc+1;         /*读数据位,接收的数据位放入retc中*/
            _nop_();
            _nop_();
        }
    scl=0;
    _nop_();
    _nop_();
    return(retc);
}


/**********************************************************
应答子函数
函数原型: void Ack_I2c(bit a);
功能:主控器进行信号应答(可以是应答或非应答,由位参数a决定)
**********************************************************/
void Ack_I2c(bit a)
{
    if(a==0) sda=0;                         /*在此发出应答或非应答信号*/
    else sda=1;                             /*0为发出应答,1为非应答信号*/
    _nop_();
    _nop_();
    _nop_();
    scl=1;
    _nop_();
    _nop_();                                /*时钟低电平周期>4μs*/
    _nop_();
    _nop_();
    _nop_();
    scl=0;                                  /*清时钟线,住I²C总线以便继续接收*/
    _nop_();
    _nop_();
}


/*********************************************************
* 函数名:Pcf8591_DaConversion
* 函数功能:PCF8591的输出端输出模拟量
* 输入:addr(器件地址)、channel(转换通道)、value(转换的数值)
* 输出:无
****************** *********************************/
bit Pcf8591_DaConversion(unsigned char addr,unsigned char channel, unsigned char Val)
{
    Start_I2c();                            // 启动总线
    I2C_SendByte(addr);                     // 发送器件地址
    if(ack==0) return(0);
    I2C_SendByte(0x40|channel);             // 发送控制字节
    if(ack==0) return(0);
```

```c
    I2C_SendByte(Val);              // 发送 DAC 的数值
    if(ack==0) return(0);
    Stop_I2c();                     // 结束总线
    return(1);
}

/******************************************************
* 函数名: Pcf8591_SendByte
* 函数功能: 写入一个控制命令
* 输入: addr(器件地址)、channel(转换通道)
* 输出: 无
******************************************************/
bit PCF8591_SendByte(unsigned char addr,unsigned char channel)
{
    Start_I2c();                    // 启动总线
    I2C_SendByte(addr);             // 发送器件地址
    if(ack==0) return(0);
    I2C_SendByte(0x40|channel);     // 发送控制字节
    if(ack==0) return(0);
    Stop_I2c();                     // 结束总线
    return(1);
}


/******************************************************
* 函数名: PCF8591_RcvByte
* 函数功能: 读取一个转换值
* 输入:
* 输出: dat
******************************************************/
unsigned char PCF8591_RcvByte(unsigned char addr)
{
    unsigned char dat;

    Start_I2c();                    // 启动总线
    I2C_SendByte(addr+1);           // 发送器件地址
    if(ack==0) return(0);
    dat=I2C_RcvByte();              // 读取数据 0

    Ack_I2c(1);                     // 发送非应答信号
    Stop_I2c();                     // 结束总线
    return(dat);
}
/*-----------------------------------------------
串口初始化函数
-----------------------------------------------*/
void init_com(void)
{
    EA=1;              // 开总中断
    ES=1;              // 允许串口中断
    ET1=1;             // 允许定时器 T1 的中断
    TMOD=0x20;         // 定时器 T1, 在方式 2 中断, 产生波特率
    PCON=0x00;         //SMOD=0
    SCON=0x50;         // 方式 1, 由定时器控制
    TH1=0xfd;          // 波特率设置为 9 600bit/s
    TL1=0xfd;
```

```c
    TR1=1;                    // 开定时器T1, 运行控制位
}
/*------------------------------------------------
延时函数
------------------------------------------------*/
void delay(unsigned char i)
{
  unsigned char j,k;
  for(j=i;j>0;j--)
    for(k=125;k>0;k--);
}
/*------------------------------------------------
把读取值转换成一个一个的字符, 送串口显示
------------------------------------------------*/
void To_ascii(unsigned char num)
{
    SBUF=num/100+'0';
    delay(200);
    SBUF=num/10%10+'0';
    delay(200);
    SBUF=num%10+'0';
    delay(200);
}
/*------------------------------------------------
主函数
------------------------------------------------*/
main()
{

  init_com();
  while(1)
  {
  /******** 以下A/D-D/A转换 *************/
   PCF8591_SendByte(AddWr,0);
   D[0]=PCF8591_RcvByte(AddWr);   //ADC0 模/数转换1    光敏电阻
  /******** 以下将AD的值通过串口发送出去 *************/
   delay(200);
   To_ascii(D[0]);
   SBUF='\n';
   delay(200);
   if(RI)
   {
     date=SBUF;      // 单片机接收
     SBUF=date;      // 单片机发送
     RI=0;
   }
   }
}
```

光敏电阻串口读取实验A/D模块实物接线图如图23-3所示。注意, 务必按图连接开发板后, 再进行整个操作。

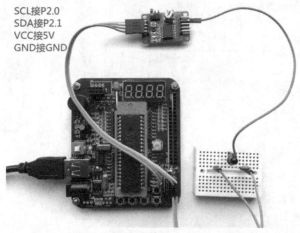

图 23-3　A/D 模块光敏电阻串口读取实验实物接线图

23.4　技术拓展

23.4.1　热感灯

热感灯实验程序如下。

```
/*******************BST-M51实验开发板例程 **********************
* 平台：BST-M51 + Keil U4 + STC89C51
* 名称：热感应灯实验
* 公司：深圳市亚博软件开发有限公司
* 日期：2015-7
* 晶振：11.059 2MHz
****************************************************************/
#include<reg52.h>        // 包含单片机寄存器的头文件
#include <intrins.h>

#define  AddWr 0x90    //PCF8591 地址
sbit beep = P2^3;
// 变量定义
unsigned char AD_CHANNEL;
unsigned long xdata   LedOut[8];
unsigned char  D[32];

sbit scl=P2^0;          //I²C 时钟
sbit sda=P2^1;          //I²C 数据
bit ack;                /*应答标志位*/
unsigned char date;
/************************************************************
启动总线函数
函数原型：void  Start_I2c();
功能：启动 I²C 总线，即发送 I²C 起始条件
**************************************************************/
void Start_I2c()
{
  sda=1;         /*发送起始条件的数据信号*/
  _nop_();
```

```
    scl=1;
    _nop_();              /* 起始条件建立时间 >4.7μs, 延时 */
    _nop_();
    _nop_();
    _nop_();
    _nop_();
    sda=0;                /* 发送起始信号 */
    _nop_();              /* 起始条件锁定时间 >4μs*/
    _nop_();
    _nop_();
    _nop_();
    _nop_();
    scl=0;                /* 钳住 I²C 总线，准备发送或接收数据 */
    _nop_();
    _nop_();
}
/******************************************************
结束总线函数
函数原型: void Stop_I2c();
功能:      结束 I²C 总线，即发送 I²C 结束条件
******************************************************/
```

23.4.2 火焰传感器报警

火焰传感器报警实验程序如下。

```
/*******************BST-M51 实验开发板例程 *********************
*  平台：BST-M51 + Keil U4 + STC89C52RC
*  名称：火焰传感器报警实验
*  晶振：11.059 2MHz
****************************************************************/
#include<reg52.h>           // 包含单片机寄存器的头文件
#include <intrins.h>

#define  AddWr 0x90         //PCF8591 地址

// 变量定义
unsigned char AD_CHANNEL;
unsigned char  D[32];

sbit scl=P2^0;              //I²C 时钟
sbit sda=P2^1;              //I²C 数据
sbit beep=P2^3;             // 蜂鸣器
bit ack;                    /* 应答标志位 */
/******************************************************
启动总线函数
函数原型: void Start_I2c();
功能：启动 I²C 总线，即发送 I²C 起始条件
******************************************************/
```

此项目附操作视频及代码资料。

Chapter 24

项目24
人体红外感应灯

项目目标

- 控制人体红外感应灯的红外智能节电开关是基于红外线技术的自动控制产品,当有人进入感应范围时,专用传感器探测到人体红外光谱的变化,自动接通负载,人不离开感应范围,将持续接通;人离开后,延时自动关闭负载,即人到灯亮,人离灯熄,亲切方便,安全节能,更显示出人性化关怀。

建议学时

- 4学时。

知识要点

- 人体红外传感器实验。
- 人体红外传感器的功能。

技能掌握

- 掌握温度数据采集的方法,以及发光电路的设计。

24.1 项目分析

红外线感应器（也称人体红外感应器）是根据红外线反射的原理研制的，属于一种智能节能设备，相关产品包括感应水龙头、自动干手器、医用洗手器、自动给皂器等，涉及领域日益广泛。

24.2 技术准备

24.2.1 基本概念及参数

人体红外感应模块即人体感应类开关又叫热释人体感应开关或红外智能开关。它是基于红外线技术的自动控制产品，当人进入感应范围时，专用传感器探测到人体红外光谱的变化，自动接通负载，人不离开感应范围，将持续接通；人离开后，延时自动关闭负载。

人体红外感应开关的主要器件为人体热释电红外传感器。人体都有恒定的体温，一般在 36~37℃，所以会发出特定波长的红外线，被动式红外探头就是用来探测人体发射的红外线的。人体发射的 9.5μm 红外线通过菲涅尔镜片增强，聚集到红外感应源上，红外感应源通常采用热释电元件，这种元件在接收到的人体红外辐射温度发生变化时，就会失去电荷平衡，向外释放电荷，后续电路经检测处理后，就能触发开关动作。人不离开感应范围，开关将持续接通；人离开后或在感应区域内长时间无动作，开关将自动延时关闭负载。HC-SR501 人体红外感应模块电气参数如表 24-1 所示。

表 24-1 HC-SR501 人体红外感应模块电气参数

产品型号	HC-SR501 人体红外感应模块
工作电压范围	直流电压 4.5~20V
静态电流	<50μA
电平输出	高 3.3V / 低 0V
触发方式	L 不可重复触发 / H 可重复触发
延时时间	0.5~200s（可调），可制作范围零点几秒~几十分钟
封锁时间	2.5s（默认），可制作范围零点几秒~几十秒
电路板外形尺寸	32mm×24mm
感应角度	<100° 锥角
工作温度	−15~+70℃
感应透镜尺寸	直径 23mm（默认）

24.2.2 功能特点

（1）全自动感应：人进入其感应范围，则输出高电平；人离开其感应范围，则自动延时关闭高电平，输出低电平。

（2）2 种触发方式（可跳线选择）如下。

① 不可重复触发方式：感应输出高电平后，延时时间段一结束，输出将自动从高电平变成低电平。

② 可重复触发方式：感应输出高电平后，在延时时间段内，如果有人体在其感应范围内活动，其

输出将一直保持高电平，直到人离开后，才延时将高电平变为低电平（感应模块检测到人体的每一次活动后，会自动顺延一个延时时间段，并且以最后一次活动的时间为延时时间的起始点）。

（3）具有感应封锁时间（默认设置为2.5s封锁时间）：感应模块在每一次感应输出后（高电平变成低电平），可以紧跟着设置一个封锁时间段，在此时间段内感应器不接受任何感应信号。此功能可以实现"感应输出时间"和"封锁时间"两者的间隔工作，可应用于间隔探测产品；同时此功能可有效抑制负载切换过程中产生的各种干扰（此时间可设置在零点几秒～几十秒）。

（4）工作电压范围宽：默认工作电压DC4.5～20V。

（5）微功耗：静态电流<50μA，特别适合干电池供电的自动控制产品。

（6）输出高电平信号：可方便与各类电路实现对接。

人体红外感应模块实物图如图24-1所示，使用时需要注意以下2点。

① 调节距离电位器，顺时针旋转，感应距离增大（最大值约7m），反之，感应距离减小（最小值约3m）。

② 调节延时电位器，顺时针旋转，感应延时加长（最大值约300s），反之，感应延时缩短（最小值约0.5s）。

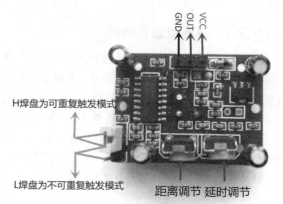

图24-1 人体红外感应模块实物图

24.2.3 使用说明

人体红外感应模块通电后有1min左右的初始化时间，在此期间模块会间隔地输出0～3次，1min后进入待机状态。

应尽量避免灯光等干扰源近距离直射模块表面的透镜，以免引进干扰信号产生误动作；使用环境尽量避免流动的风，风也会对感应器造成干扰。

感应模块采用双元探头，探头的窗口为长方形，双元（A元B元）位于较长方向的两端，当人体从左到右或从右到左走过时，红外光谱到达双元的时间、距离有差值，差值越大，感应越灵敏；当人体从正面走向探头或从上到下或从下到上方向走过时，双元检测不到红外光谱距离的变化，无差值，因此感应不灵敏或不工作。所以安装感应器时应使探头双元的方向与人体活动最多的方向尽量相平行，保证人体经过时先后被探头双元所感应。为了增加感应角度范围，本模块采用圆形透镜，也使得探头四面都感应，但左右两侧仍然比上下两个方向感应范围大、灵敏度强，仍须尽量按以上要求安装。人体红外感应模块感应范围如图24-2所示。

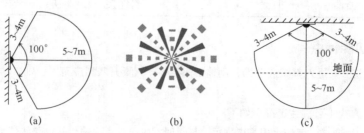

图24-2 人体红外感应模块感应范围图示

24.3 项目实施

串口读取电压值实验程序如下。

```c
/********************BST-M51 实验开发板例程 ********************
* 平台:BST-M51 + Keil U4 + STC89C51
* 名称:LCD1602 模块实验
* 晶振:11.059 2MHz
*************************************************************/
#include<reg52.h>      //包含单片机寄存器的头文件
#include <intrins.h>

#define  AddWr 0x90     //PCF8591 地址

// 变量定义
unsigned char AD_CHANNEL=0;
unsigned char  D[32];

sbit scl=P2^0;         //I²C 时钟
sbit sda=P2^1;         //I²C 数据
bit ack;                /*应答标志位 */

unsigned char date;

/*************************************************************
启动总线函数
函数原型: void  Start_I2c();
功能:启动 I²C 总线,即发送 I²C 起始条件
*************************************************************/
void Start_I2c()
{
sda=1;                /*发送起始条件数据信号 */
  _nop_();
scl=1;
  _nop_();            /*起始条件建立时间>4.7μs,延时 */
  _nop_();
  _nop_();
  _nop_();
  _nop_();
sda=0;                /*发送起始信号 */
  _nop_();            /* 起始条件锁定时间>4μs*/
  _nop_();
  _nop_();
  _nop_();
  _nop_();
scl=0;                /*钳住 I²C 总线,准备发送或接收数据 */
  _nop_();
  _nop_();
}

/*************************************************************
结束总线函数
函数原型: void  Stop_I2c();
功能:结束 I²C 总线,即发送 I²C 结束条件
*************************************************************/
void Stop_I2c()
```

```
    {
    sda=0;              /*发送结束条件的数据信号*/
     _nop_();           /*发送结束条件的时钟信号*/
    scl=1;              /*结束条件建立时间>4μs*/
    _nop_();
     _nop_();
     _nop_();
     _nop_();
    sda=1;          /*发送I²C总线结束信号*/
     _nop_();
     _nop_();
     _nop_();
     _nop_();
    }


/*****************************************************************
字节数据发送函数
函数原型: void I2C_SendByte (UCHAR c);
功能:将数据c发送出去,可以是地址,也可以是数据,发送完毕后等待应答,并对此状态位进行操作,ack=1,发送数据
正常; ack=0,被控器件无应答或损坏(即不应答,或非应答,都使ack=0)
*****************************************************************/
    void I2C_SendByte(unsigned char c)
    {
     unsigned char  i;

      for(i=0;i<8;i++) /*要传送的数据长度为8位*/
         {
          if((c<<i)&0x80) sda=1;/*判断发送位*/
             else   sda=0;
          _nop_();
          scl=1;                /*置时钟线为高电平,通知被控器开始接收数据位*/
          _nop_();
          _nop_();              /*保证时钟高电平周期>4μs*/
          _nop_();
          _nop_();
          _nop_();
          scl=0;
          }

      _nop_();
_nop_();
      sda=1;            /*8位发送完后,释放数据线,准备接收应答位*/
_nop_();
       _nop_();
       scl=1;
       _nop_();
       _nop_();
       _nop_();
       if(sda==1) ack=0;
         else ack=1;    /*判断是否接收到应答信号*/
       scl=0;
       _nop_();
       _nop_();
     }

/*****************************************************************
```

字节数据接收函数
函数原型：UCHAR I2C_RcvByte();
功能：用来接收从器件传来的数据，并判断总线错误（不发应答信号），发完后请用应答函数应答从机
***/
```c
unsigned char  I2C_RcvByte()
{
  unsigned char  retc=0,i;
  sda=1;                            /*置数据线为输入方式*/
  for(i=0;i<8;i++)
    {
      _nop_();
      scl=0;                        /*置时钟线为低电平，准备接收数据位*/
      _nop_();
      _nop_();                      /*时钟低电平周期>4.7μs*/
      _nop_();
      _nop_();
      _nop_();
      scl=1;                        /*置时钟线为高电平，使数据线上数据有效*/
      _nop_();
      _nop_();
      retc=retc<<1;
      if(sda==1) retc=retc+1;       /*读数据位，接收的数据位放入 retc 中 */
      _nop_();
      _nop_();
    }
  scl=0;
  _nop_();
  _nop_();
  return(retc);
}


/*****************************************************************
应答子函数
函数原型： void Ack_I2c(bit a);
功能：主控器进行信号应答（可以是应答或非应答，由位参数 a 决定）
*****************************************************************/
void Ack_I2c(bit a)
{
  if(a==0) sda=0;                   /*在此发出应答或非应答信号*/
  else sda=1;                       /*0 为发出应答，1 为非应答信号*/
  _nop_();
  _nop_();
  _nop_();
  scl=1;
  _nop_();
  _nop_();                          /*时钟低电平周期>4μs*/
  _nop_();
  _nop_();
  _nop_();
  scl=0;                            /*清时钟线，钳住 I²C 总线，以便继续接收*/
  _nop_();
  _nop_();
}


/*****************************************************
* 函数名:Pcf8591_DaConversion
* 函数功能：PCF8591 的输出端输出模拟量
```

* 输入：addr（器件地址）、channel（转换通道）、value（转换的数值）
* 输出：无
***/
```c
bit Pcf8591_DaConversion (unsigned char addr,unsigned char channel,  unsigned char Val)
{
    Start_I2c();                        // 启动总线
    I2C_SendByte(addr);                 // 发送器件地址
    if(ack==0)return(0);
    I2C_SendByte(0x40|channel);         // 发送控制字节
    if(ack==0)return(0);
    I2C_SendByte(Val);                  // 发送DAC的数值
    if(ack==0)return(0);
    Stop_I2c();                         // 结束总线
    return(1);
}
```

/***
* 函数名：Pcf8591_SendByte
* 函数功能：写入一个控制命令
* 输入：addr（器件地址）、channel（转换通道）
* 输出：无
***/
```c
bit PCF8591_SendByte (unsigned char addr,unsigned char channel)
{
    Start_I2c();                        // 启动总线
    I2C_SendByte(addr);                 // 发送器件地址
    if(ack==0)return(0);
    I2C_SendByte(0x40|channel);         // 发送控制字节
    if(ack==0)return(0);
    Stop_I2c();                         // 结束总线
    return(1);
}
```

/***
* 函数名：PCF8591_RcvByte
* 函数功能：读取一个转换值
* 输入：
* 输出：dat
***/
```c
unsigned char PCF8591_RcvByte (unsigned char addr)
{
    unsigned char dat;

    Start_I2c();                        // 启动总线
    I2C_SendByte(addr+1);               // 发送器件地址
    if(ack==0)return(0);
    dat=I2C_RcvByte();                  // 读取数据0

    Ack_I2c(1);                         // 发送非应答信号
    Stop_I2c();                         // 结束总线
    return(dat);
}
```
/*---
串口初始化函数
---*/
```c
void init_com (void)
{
```

```c
    EA=1;          // 开总中断
    ES=1;          // 允许串口中断
    ET1=1;         // 允许定时器 T1 的中断
    TMOD=0x20;     // 定时器 T1，在方式 2 中断产生波特率
    PCON=0x00;     //SMOD=0
    SCON=0x50;     // 方式 1 由定时器控制
    TH1=0xfd;      // 波特率设置为 9 600bit/s
    TL1=0xfd;
    TR1=1;         // 开定时器 T1，运行控制位

}
/*------------------------------------------------
延时函数
------------------------------------------------*/
void delay(unsigned char i)
{
  unsigned char j,k;
  for(j=i;j>0;j--)
    for(k=125;k>0;k--);
}
/*------------------------------------------------
把读取值转换成一个一个的字符，送串口显示
------------------------------------------------*/
void To_ascii(unsigned char num)
{
    SBUF=num/100+'0';
    delay(200);
    SBUF=num/10%10+'0';
    delay(200);
    SBUF=num%10+'0';
    delay(200);
}
/*------------------------------------------------
主函数
------------------------------------------------*/
main()
{

  init_com();
  while(1)
  {
  /******** 以下 A/D-D/A 转换 *************/
  PCF8591_SendByte(AddWr,2);
  D[2]=PCF8591_RcvByte(AddWr);
  /******** 以下将 AD 的值通过串口发送出去 *************/
  delay(200);
   To_ascii(D[2]);
   SBUF='\n';
  delay(200);
   if(RI)
   {
    date=SBUF;     // 单片机接收
    SBUF=date;     // 单片机发送
    RI=0;
   }
  }
}
```

串口读取电压值实验 LCD1620 模块实物接线图如图 24-3 所示。
注意，务必按图连接开发板，然后再进行整个操作。

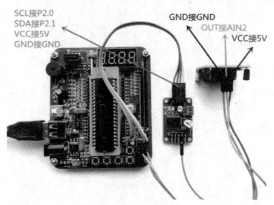

图 24-3　LCD1620 模块串口读取电压值实验实物接线图

24.4　技术拓展

人体红外感应灯实验程序如下。

```
/********************BST-M51 实验开发板例程 *********************
* 平台: BST-M51 + Keil U4 + STC89C51
* 名称: 人体红外感应灯实验
* 晶振: 11.059 2MHz
***************************************************************/
#include<reg52.h>        // 包含单片机寄存器的头文件
#include <intrins.h>

#define   AddWr 0x90      //PCF8591 地址

// 变量定义
unsigned char AD_CHANNEL=0;
unsigned char   D[32];

sbit scl=P2^0;            //I²C 时钟
sbit sda=P2^1;            //I²C 数据
bit ack;                   /* 应答标志位 */
unsigned char date;
unsigned int time_count=0;
bit flag=0;
/***************************************************************
启动总线函数
函数原型 : void  Start_I2c();
功能：启动 I²C 总线，即发送 I²C 起始条件
***************************************************************/
```

此项目附操作视频及代码资料。

Chapter 25

项目25
无线模块

项目目标

- 无线模块广泛地运用在通信控制、工业自动化、智能家居、生物信号采集、机器人控制等领域中。在无线模块信号发射、接收实验的操作过程中,读者要了解无线模块的硬件结构和程序编制,并不断深化。

建议学时

- 4学时。

知识要点

- 无线模块开发流程。
- 无线模块程序烧录。

技能掌握

- 了解NRF24L01的原理和工作模式,熟练掌握C语言的时序函数。

25.1 项目分析

无线模块功能分为数据发送和数据接收，本项目通过学习无线模块 NRF24L01 的基本编程、接口的链接，初步了解无线模块的实际用途。

25.2 技术准备

25.2.1 NRF24L01 简介

NRF24L01 是由 NORDIC 生产的，工作在 2.4 ～ 2.5GHz ISM 频段的单片无线收发器芯片。输出功率频道选择和协议可以通过 SPI 端口进行设置。几乎可以和各种单片机芯片连接，并完成无线数据传送工作。电流消耗极低，当工作在发射模式下，发射功率为 0dBm 时，电流消耗为 11.3mA；接收模式时为 12.3mA；掉电模式和待机模式下电流消耗更低。

25.2.2 模块外接引脚

BST-M51 原理图如图 25-1 所示。

CSN（即图中 CS）：芯片的片选线，CSN 为低电平芯片工作。

SCK：芯片控制的时钟（SPI 时钟）线。

MISO（即图中 MIS）：SPI 主机输入从机输出端（Master Input Slave Output）。

MOSI（即为图中 MOS）：SPI 主机输出从机输入端（Master Output Slave Input） IRQ：中断信号。引脚会在以下 3 种情况下变低。

① Tx FIFO 发完并且收到 ACK（使能 ACK 情况下）。

② Rx FIFO 收到数据。

③ 达到最大重发次数。

CE（即图中 RST）：芯片的模式控制线。在 CSN 为低电平的情况下，CE 协同 NRF24L01 的 CONFIG 寄存器共同决定 NRF24L01 的状态。

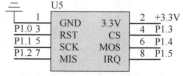

图 25-1　BST-M51 原理图

25.2.3 SPI

SPI 即串行外设接口，是一种高速、全双工、同步的通信总线协议。

SPI 指令设置，CSN 为低电平后，SPI 接口等待执行指令。每一条指令的执行都必须通过 1 次 CSN 由高到低的变化。SPI 读操作时序图如图 25-2（a）所示，SPI 写操作时序图如图 25-2（b）所示。关于这部分的几个函数如下。

1. 基本函数 uchar SPI_RW（uchar byte）

```
uchar SPI_RW (uchar byte)
{
uchar bit_ctr;
for (bit_ctr=0;bit_ctr<8;bit_ctr++) // output 8-bit
{
MOSI = (byte & 0x80); // output 'byte', MSB to MOSI
byte = (byte << 1); // shift next bit into MSB..
SCK = 1; // Set SCK high..
byte |= MISO; // capture current MISO bit
SCK = 0; // ..then set SCK low again
```

```
}
return(byte); // return read byte
}
```

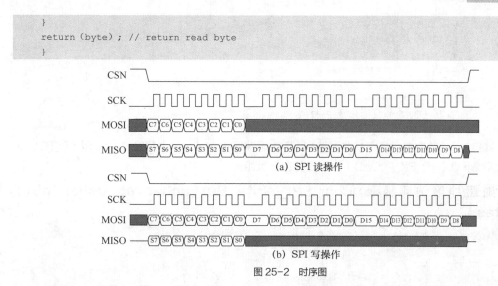

图 25-2 时序图

最基本的函数用于完成 GPIO 模拟 SPI 的功能。将输出字节（MOSI）从 MSB 循环输出，同时将输入字节（MISO）从 LSB 循环移入。上升沿读入，下降沿输出（从 SCK 被初始化为低电平可以做出判断）。

2. 寄存器访问函数 uchar SPI_RW_Reg（uchar reg, uchar value）

```
uchar SPI_RW_Reg(uchar reg, uchar value)
{
uchar status;
CSN = 0; // CSN low, init SPI transaction
status = SPI_RW(reg); // select register
SPI_RW(value); // ..and write value to it..
CSN = 1; // CSN high again
return(status); // return nRF24L01 status byte
}
```

寄存器访问函数用来设置 NRF24L01 的寄存器的值。基本思路就是通过 WRITE_REG 命令（即 0x20+ 寄存器地址）把要设定的值写到相应的寄存器地址里面去，并读取返回值。对于函数来说，也就是把 value 值写到 reg 寄存器中。

需要注意的是，访问 NRF24L01 之前首先要启动芯片（CSN=0;），访问完毕后，再禁用芯片（CSN=1;）。

3. 读取寄存器值函数 uchar SPI_Read（uchar reg）

```
uchar SPI_Read(uchar reg)
{
uchar reg_val;
CSN = 0; // CSN low, initialize SPI communication...
SPI_RW(reg); // Select register to read from..
reg_val = SPI_RW(0); // ..then read registervalue
CSN = 1; // CSN high, terminate SPI communication
return(reg_val); // return register value
}
```

读取寄存器值函数的基本思路就是通过 READ_REG 命令（即 0x00+ 寄存器地址），把寄存器中的值读出来。对于函数来说，也就是把 reg 寄存器的值读到 reg_val 中去。

4. 接收缓冲区访问函数 uchar SPI_Read_Buf（uchar reg, uchar *pBuf, uchar bytes）

```
uchar SPI_Read_Buf(uchar reg, uchar *pBuf, uchar bytes)
{
```

```c
uchar status,byte_ctr;
CSN = 0; // Set CSN low, init SPI tranaction
status = SPI_RW(reg); // Select register to write to and read status byte
for(byte_ctr=0;byte_ctr<bytes;byte_ctr++)
pBuf[byte_ctr] = SPI_RW(0); // Perform SPI_RW to read byte from nRF24L01
CSN = 1; // Set CSN high again
return(status); // return nRF24L01 status byte
}
```

接收缓冲区访问函数主要用来在接收时读取 FIFO 缓冲区中的值。基本思路是通过 READ_REG 命令把数据从接收 FIFO（RD_RX_PLOAD）中读出并存入数组。

5. 发射缓冲区访问函数 uchar SPI_Write_Buf（uchar reg, uchar *pBuf, uchar bytes）

```c
uchar SPI_Write_Buf(uchar reg, uchar *pBuf, uchar bytes)
{
uchar status,byte_ctr;
CSN = 0; // Set CSN low, init SPI tranaction
status = SPI_RW(reg); // Select register to write to and read status byte
Uart_Delay(10);
for(byte_ctr=0; byte_ctr<bytes; byte_ctr++) // then write all byte in buffer(*pBuf)
SPI_RW(*pBuf++);
CSN = 1; // Set CSN high again
return(status); // return nRF24L01 status byte
}
```

发射缓冲区访问函数主要用来把数组里的数放到发射 FIFO 缓冲区中。基本思路就是通过 WRITE_REG 命令把数据存到发射 FIFO（WR_TX_PLOAD）中去。

25.2.4　工作模式

NRF242L01 工作模式见表 25-1，引脚功能见表 25-2。

表 25-1　NRF24L01 工作模式

模　　式	PWR_UP	PRIM_RX	CE	FIFO 寄存器状态
接收模式	1	1	1	—
发送模式	1	0	1	数据在 TX FIFO 寄存器中
待机模式 2	1	0	1→0	停留在发送模式，直至数据发送完
待机模式 1	1	0	1	TX FIFO 为空
断电模式	0	—	—	—

表 25-2　NRF24L01 引脚功能

引 脚 名 称	方向	发送模式	接收模式	待机模式	断电模式
CE	输入	高电平 >10μs	高电平	低电平	—
CSN	输入	SPI 片选使能，低电平使能			
SCK	输入	SPI 时钟			
MOSI	输入	SPI 串行输入			
MISO	三态输出	SPI 串行输出			
IRQ	输出	中断，低电平使能			

NRF24L01 的 2 种常用模式程序如下。

1. 接收模式

```
void RX_Mode(void)
{
CE=0;
SPI_Write_Buf(WRITE_REG + RX_ADDR_P0, TX_ADDRESS, TX_ADR_WIDTH);//写 Rx 结点的地址
SPI_RW_Reg(WRITE_REG + EN_AA, 0x01); // 使能 AUTO ACK
SPI_RW_Reg(WRITE_REG + EN_RXADDR, 0x01); // 使能 PIPE0 SPI_RW_Reg(WRITE_REG + RF_CH, 40);
//选择通信频率
SPI_RW_Reg(WRITE_REG + RX_PW_P0, TX_PLOAD_WIDTH);//选择通道 0 有效数据
SPI_RW_Reg(WRITE_REG + RF_SETUP, 0x07);//配置发送参数（低噪放大器增益、发射功率、无线速率）
SPI_RW_Reg(WRITE_REG + CONFIG, 0x0f); // 配置 NRF24L01 的基本参数以及切换工作模式
CE = 1; // Set CE pin high to enable RX device
```

2. 发送模式

```
void TX_Mode(void)
{
CE=0;
SPI_Write_Buf(WRITE_REG + TX_ADDR, TX_ADDRESS, TX_ADR_WIDTH);//写 Tx 结点的地址
SPI_Write_Buf(WRITE_REG + RX_ADDR_P0, TX_ADDRESS, TX_ADR_WIDTH);//写 Rx 结点的地址（主要是为了使能 Auto Ack）
SPI_Write_Buf(WR_TX_PLOAD, tx_buf, TX_PLOAD_WIDTH); //选择通道 0 有效数据宽度
SPI_RW_Reg(WRITE_REG + EN_AA, 0x01); // 使能 AUTO ACK
SPI_RW_Reg(WRITE_REG + EN_RXADDR, 0x01); // 使能 PIPE 0
SPI_RW_Reg(WRITE_REG + SETUP_RETR, 0x1a);//配置自动重发次数
SPI_RW_Reg(WRITE_REG + RF_CH, 40);//选择通信频率
SPI_RW_Reg(WRITE_REG + RF_SETUP, 0x07);//配置发送参数
SPI_RW_Reg(WRITE_REG + CONFIG, 0x0e);//配置 NRF24L01 的基本参数以及切换工作模式
CE=1;
}
```

25.2.5 数据通道

NRF24L01 配置为接收模式时，可以接收 6 路不同地址相同频率的数据。每个数据通道拥有自己的地址，并且可以通过寄存器分别进行配置。数据通道 0 有 40 位可配置地址。数据通道 1 ~ 5 的地址为，32 位共用地址 + 各自的地址（最低字节）。图 25-3 所示是数据通道 1 ~ 5 的地址设置方法举例。所有数据通道可以设置多达 40 位，但是 1 ~ 5 数据通道的最低位必须不同。

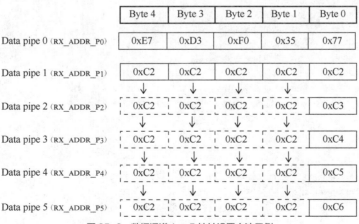

图 25-3　数据通道 1 ~ 5 地址设置方法示例

25.3 项目实施

NRF24L01一对一无线发送模块实验程序如下。

```c
/*******************BST-M51实验开发板例程************************
* 平台:BST-M51 + Keil U4 + STC89C51
* 名称:NRF24L01无线发送实验
* 晶振:11.059 2MHz
***************************************************************/
#include "reg52.h"
#include "nrf24l01.h"
unsigned char code table[]={0x3f,0x06,0x5b,0x4f,0x66,0x6d,0x7d,
                            0x07,0x7f,0x6f,0x77,0x7c,0x39,0x5e,0x79,0x71};
/******************* 主函数 **************************/
void main()
{
   unsigned int Date;
   char TxDate[4];
   NRF24L01Int();
   while(1)
    {
     TxDate[0]=table[1];
     TxDate[1]=table[5];
     TxDate[2]=table[6];
     TxDate[3]=table[8];
     NRFSetTxMode(TxDate);
     while(CheckACK());      //检测是否发送完毕
    }


}


/*******************BST-M51实验开发板例程************************
* 平台:BST-M51 + Keil U4 + STC89C51
* 名称:NRF24L01无线发送实验
* 晶振:11.059 2MHz
***************************************************************/
#include "reg52.h"
#include "nrf24l01.h"
sbit CE=P1^0;     //RX/TX模式选择端
sbit SCLK=P1^1;   //SPI时钟端
sbit MISO=P1^2;   //SPI主机输入从机输出端
sbit CSN=P1^3;  //SPI片选端 // 就是SS
sbit MOSI=P1^4;   //SPI主机输出从机输入端
sbit IRQ=P1^5;  // 可屏蔽中断端
unsigned char code TxAddr[]={0x34,0x43,0x10,0x10,0x01};// 发送地址
/*************** 状态标志 ****************************/
unsigned char bdata sta;    // 状态标志
sbit RX_DR=sta^6;
sbit TX_DS=sta^5;
sbit MAX_RT=sta^4;
/******************** 延时函数 ***************/
void Delay(unsigned int t)
{
  unsigned int x,y;
```

```c
    for(x=t;x>0;x--)
     for(y=110;y>0;y--);
}
/****************SPI 时序函数 *****************************************/
unsigned char NRFSPI(unsigned char date)
{
   unsigned char i;
   for(i=0;i<8;i++)           // 循环 8 次
   {
     if(date&0x80)
       MOSI=1;
     else
       MOSI=0;      // byte 最高位输出到 MOSI
     date<<=1;                  // 低一位移位到最高位
     SCLK=1;
     if(MISO)                   // 拉高 SCK 电平,NRF24L01 从 MOSI 读入 1 位数据,同时从 MISO 输出 1 位数据
       date|=0x01;              // 读 MISO 到 byte 最低位
     SCLK=0;                    // SCK 置低电平
   }
   return(date);                // 返回读出的 1 字节
}
/******************NRF24L01 初始化函数 ******************************/
void NRF24L01Int()
{
  Delay(2);//让系统什么都不干
  CE=0;  //待机模式 1
  CSN=1;
  SCLK=0;
  IRQ=1;
}
/****************SPI 读寄存器 1 字节函数 *****************************/
unsigned char NRFReadReg(unsigned char RegAddr)
{
   unsigned char BackDate;
   CSN=0;//启动时序
   NRFSPI(RegAddr);//写寄存器地址
   BackDate=NRFSPI(0x00);//写入读寄存器指令
   CSN=1;
   return(BackDate);  //返回状态
}
/****************SPI 写寄存器 1 字节函数 *****************************/
unsigned char NRFWriteReg(unsigned char RegAddr,unsigned char date)
{
   unsigned char BackDate;
   CSN=0;//启动时序
   BackDate=NRFSPI(RegAddr);//写入地址
   NRFSPI(date);//写入值
   CSN=1;
   return(BackDate);
}
/****************SPI 读取 RXFIFO 寄存器的值 *****************************/
unsigned char NRFReadRxDate(unsigned char RegAddr,unsigned char *RxDate,unsigned char DateLen)
{   //寄存器地址 //读取数据存放变量 //读取数据长度 //用于接收
   unsigned char BackDate,i;
  CSN=0;//启动时序
  BackDate=NRFSPI(RegAddr);//写入要读取的寄存器地址
```

```c
        for(i=0;i<DateLen;i++) // 读取数据
        {
            RxDate[i]=NRFSPI(0);
        }
        CSN=1;
        return(BackDate);
}
/***************SPI 写入 TXFIFO 寄存器的值 ******************************/
unsigned char NRFWriteTxDate(unsigned char RegAddr,unsigned char *TxDate,unsigned char DateLen)
{   // 寄存器地址 // 写入数据存放变量 // 读取数据长度 // 用于发送
    unsigned char BackDate,i;
    CSN=0;
    BackDate=NRFSPI(RegAddr);// 写入要写入寄存器的地址
    for(i=0;i<DateLen;i++) // 写入数据
    {
        NRFSPI(*TxDate++);
    }
    CSN=1;
    return(BackDate);
}
/****************NRF 设置为发送模式并发送数据 ***************************/
void NRFSetTxMode(unsigned char *TxDate)
{// 发送模式
    CE=0;
    NRFWriteTxDate(W_REGISTER+TX_ADDR,TxAddr,TX_ADDR_WITDH);// 写寄存器指令 + 接收地址使能指令 + 接收地址 + 地址宽度
    NRFWriteTxDate(W_REGISTER+RX_ADDR_P0,TxAddr,TX_ADDR_WITDH);// 为了应答接收设备，接收通道 0 地址和发送地址相同
    NRFWriteTxDate(W_TX_PAYLOAD,TxDate,TX_DATA_WITDH);// 写入数据
    /***** 下面有关寄存器配置 *************/
    NRFWriteReg(W_REGISTER+EN_AA,0x01);          // 使能接收通道 0 自动应答
    NRFWriteReg(W_REGISTER+EN_RXADDR,0x01);      // 使能接收通道 0
    NRFWriteReg(W_REGISTER+SETUP_RETR,0x0a);     // 自动重发延时等待 250μs+86μs,自动重发 10 次
    NRFWriteReg(W_REGISTER+RF_CH,0x40);          // 选择射频通道 0x40
    NRFWriteReg(W_REGISTER+RF_SETUP,0x07);       // 数据传输率 1Mbit/s,发射功率 0dBm,低噪声放大器增益
    NRFWriteReg(W_REGISTER+CONFIG,0x0e);         // CRC 使能，16 位 CRC 校验，通电
    CE=1;
    Delay(5);// 保持 10μs 秒以上
}
/****************NRF 设置为接收模式并接收数据 ***************************/
// 主要接收模式
void NRFSetRXMode()
{
    CE=0;
    NRFWriteTxDate(W_REGISTER+RX_ADDR_P0,TxAddr,TX_ADDR_WITDH);   // 接收设备接收通道 0 使用和发送设备相同的发送地址
    NRFWriteReg(W_REGISTER+EN_AA,0x01);                  // 使能接收通道 0 自动应答
    NRFWriteReg(W_REGISTER+EN_RXADDR,0x01);              // 使能接收通道 0
    NRFWriteReg(W_REGISTER+RF_CH,0x40);                  // 选择射频通道 0x40
    NRFWriteReg(W_REGISTER+RX_PW_P0,TX_DATA_WITDH);      // 接收通道 0 选择和发送通道相同有效数据宽度
    NRFWriteReg(W_REGISTER+RF_SETUP,0x07);               // 数据传输率 1Mbit/s,发射功率 0dBm,低噪声放大器增益 */
    NRFWriteReg(W_REGISTER+CONFIG,0x0f);                 // CRC 使能，16 位 CRC 校验，通电，接收模式
    CE = 1;
    Delay(5);// 保持 10μs 秒以上
```

```c
}
/*********************** 检测应答信号 *****************************/
unsigned char CheckACK()
{  // 用于发送
  sta=NRFReadReg(R_REGISTER+STATUS);           // 返回状态寄存器
  if(TX_DS||MAX_RT)// 发送完毕中断
  {
     NRFWriteReg(W_REGISTER+STATUS,0xff);      // 清除 TX_DS 或 MAX_RT 中断标志
     CSN=0;
     NRFSPI(FLUSH_TX);// 用于清空 FIFO, 非常关键, 否则会出现意想不到的后果
CSN=1;
  return(0);
  }
  else
  return(1);
}
/**************** 判断是否接收到数据, 接收到就从 RX 取出 ********************/
// 用于接收模式
unsigned char NRFRevDate(unsigned char *RevDate)
{
   unsigned char RevFlags=0;
   sta=NRFReadReg(R_REGISTER+STATUS);// 发送数据后读取状态寄存器
    if(RX_DR)                         // 判断是否接收到数据
    {
       CE=0;                          //SPI 使能
    NRFReadRxDate(R_RX_PAYLOAD,RevDate,RX_DATA_WITDH);// 从 RXFIFO 读取数据
    RevFlags=1;        // 读取数据完成标志
     }
   NRFWriteReg(W_REGISTER+STATUS,0xff); // 接收到数据后 RX_DR,TX_DS,MAX_PT 都置 1, 通过写 1 来清除中断标志
   return(RevFlags);
}


#include "reg52.h"
#include "nrf24l01.h"
#include "delay.h"
/********************** 延时函数 ***************/
void Delay(unsigned int t)
{
  unsigned int x,y;
  for(x=t;x>0;x--)
   for(y=110;y>0;y--);
}
/*******************************************
   DS18B20 专用延时子函数
*******************************************/
void DDelay(unsigned int t)
{
  unsigned int n;
  n=0;
  while(n<t)
  {
    n++;
  }
}
```

（2）NRF24L01 一对一无线接收模块实验程序如下。

```c
/******************BST-M51 实验开发板例程 ************************
*   平台：BST-M51 + Keil U4 + STC89C51
*   名称：24L01 无线发送实验
*   晶振：11.059 2MHz
*********************************************************/
#include "reg52.h"
#include "nrf24l01.h"
/****************** 主函数 ********************/
void main()
{
    NRF24L01Int();
    while(1)
    {
        NRFSetRXMode();// 设置为接收模式
        GetDate();// 开始接收数
    }
}


/******************BST-M51 实验开发板例程 ************************
*   平台：BST-M51 + Keil U4 + STC89C51
*   名称：24L01 无线接收实验
*   晶振：11.059 2MHz
*********************************************************/
#include "reg52.h"
#include "nrf24l01.h"
sbit CE=P1^0;    //RX/TX 模式选择端
sbit SCLK=P1^1;     //SPI 时钟端
sbit MISO=P1^2;     //SPI 主机输入从机输出端
sbit CSN=P1^3;  //SPI 片选端 // 就是 SS
sbit MOSI=P1^4;     //SPI 主机输出从机输入端
sbit IRQ=P1^5;  // 可屏蔽中断端
sbit wei1=P2^4;
sbit wei2=P2^5;
sbit wei3=P2^6;
sbit wei4=P2^7;
unsigned char Date[5];// 最后一位用来存放结束标志
unsigned char code TxAddr[]={0x34,0x43,0x10,0x10,0x01};// 发送地址
/*************** 状态标志 *****************************/
unsigned char  bdata sta;    // 状态标志
sbit RX_DR=sta^6;
sbit TX_DS=sta^5;
sbit MAX_RT=sta^4;

/******************** 延时函数 **********************/
void NRFDelay(unsigned int t)
{
    unsigned int x,y;
    for(x=t;x>0;x--)
     for(y=110;y>0;y--);
}
/*****************LED 显示器显示 ********************/
unsigned char Shumaguan (unsigned char *x)
{
        wei1=1;
```

```c
        wei2=0;
        wei3=0;
        wei4=0;
        P0=x[0];
        NRFDelay(2);
        wei1=0;
        wei2=1;
        wei3=0;
        wei4=0;
        P0=x[1];
        NRFDelay(2);
        wei1=0;
        wei2=0;
        wei3=1;
        wei4=0;
        P0=x[2];
        NRFDelay(2);
        wei1=0;
        wei2=0;
        wei3=0;
        wei4=1;
        P0=x[3];
        NRFDelay(2);
}
/****************SPI 时序函数 *****************************************/
unsigned char NRFSPI(unsigned char date)
{
    unsigned char i;
    for(i=0;i<8;i++)              // 循环8次
    {
        if(date&0x80)
            MOSI=1;
        else
            MOSI=0;               // byte 最高位输出到 MOSI
        date<<=1;                 // 低一位移位到最高位
        SCLK=1;
        if(MISO)                  // 拉高 SCK 电平,NRF24L01 从 MOSI 读入1位数据,同时从 MISO 输出1位数据
            date|=0x01;           // 读 MISO 到 byte 最低位
        SCLK=0;                   // SCK 置低电平
    }
    return(date);                 // 返回读出的1字节
}
/*******************NRF24L01 初始化函数 ****************************/
void NRF24L01Int()
{
    NRFDelay(2);//让系统什么都不干
    CE=0;
    CSN=1;
    SCLK=0;
    IRQ=1;
}
/****************SPI 读寄存器1字节函数 ****************************/
unsigned char NRFReadReg(unsigned char RegAddr)
{
    unsigned char BackDate;
    CSN=0;//启动时序
```

```c
    NRFSPI(RegAddr);// 写寄存器地址
    BackDate=NRFSPI(0x00);// 写入读寄存器指令
    CSN=1;
    return(BackDate); // 返回状态
}
/****************SPI 写寄存器 1 字节函数 *******************************/
unsigned char NRFWriteReg(unsigned char RegAddr,unsigned char date)
{
    unsigned char BackDate;
    CSN=0;// 启动时序
    BackDate=NRFSPI(RegAddr);// 写入地址
    NRFSPI(date);// 写入值
    CSN=1;
    return(BackDate);
}
/****************SPI 读取 RXFIFO 寄存器的值 *******************************/
unsigned char NRFReadRxDate(unsigned char RegAddr,unsigned char *RxDate,unsigned char DateLen)
{   // 寄存器地址 // 读取数据存放变量 // 读取数据长度 // 用于接收
    unsigned char BackDate,i;
    CSN=0;// 启动时序
    BackDate=NRFSPI(RegAddr);// 写入要读取的寄存器地址
    for(i=0;i<DateLen;i++)// 读取数据
    {
        RxDate[i]=NRFSPI(0);
    }
    CSN=1;
    return(BackDate);
}
/****************SPI 写入 TXFIFO 寄存器的值 *******************************/
unsigned char NRFWriteTxDate(unsigned char RegAddr,unsigned char *TxDate,unsigned char DateLen)
{   // 寄存器地址 // 写入数据存放变量 // 读取数据长度 // 用于发送
    unsigned char BackDate,i;
    CSN=0;
    BackDate=NRFSPI(RegAddr);// 写入要写入寄存器的地址
    for(i=0;i<DateLen;i++)// 写入数据
    {
        NRFSPI(*TxDate++);
    }
    CSN=1;
    return(BackDate);
}
/****************NRF 设置为发送模式并发送数据 *******************************/
void NRFSetTxMode(unsigned char *TxDate)
{   // 发送模式
    CE=0;
    NRFWriteTxDate(W_REGISTER+TX_ADDR,TxAddr,TX_ADDR_WITDH);// 写寄存器指令 +P0 地址使能指令 + 发送地址 + 地址宽度
    NRFWriteTxDate(W_REGISTER+RX_ADDR_P0,TxAddr,TX_ADDR_WITDH);// 为了应答接收设备，接收通道 0 地址和发送地址相同
    NRFWriteTxDate(W_TX_PAYLOAD,TxDate,TX_DATA_WITDH);// 写入数据
    /****** 下面有关寄存器配置 **************/
    NRFWriteReg(W_REGISTER+EN_AA,0x01);         // 使能接收通道 0 自动应答
    NRFWriteReg(W_REGISTER+EN_RXADDR,0x01);     // 使能接收通道 0
    NRFWriteReg(W_REGISTER+SETUP_RETR,0x0a);    // 自动重发延时等待 250μs+86μs，自动重发 10 次
    NRFWriteReg(W_REGISTER+RF_CH,0x40);         // 选择射频通道 0x40
    NRFWriteReg(W_REGISTER+RF_SETUP,0x07);      // 数据传输率 1Mbit/s，发射功率 0dBm，低噪声放大器增益
```

```c
    NRFWriteReg(W_REGISTER+CONFIG,0x0e);          // CRC 使能，16 位 CRC 校验，通电
    CE=1;
    NRFDelay(5);//保持 10μs 秒以上
}
/*****************NRF 设置为接收模式并接收数据 *****************************/
//接收模式
void NRFSetRXMode()
{
    CE=0;
    NRFWriteTxDate(W_REGISTER+RX_ADDR_P0,TxAddr,TX_ADDR_WITDH);   // 接收设备接收通道 0 使用和发送设备相同的发送地址
    NRFWriteReg(W_REGISTER+EN_AA,0x01);              // 使能接收通道 0 自动应答
    NRFWriteReg(W_REGISTER+EN_RXADDR,0x01);          // 使能接收通道 0
    NRFWriteReg(W_REGISTER+RF_CH,0x40);              // 选择射频通道 0x40
    NRFWriteReg(W_REGISTER+RX_PW_P0,TX_DATA_WITDH);  // 接收通道 0 选择和发送通道相同有效数据宽度
    NRFWriteReg(W_REGISTER+RF_SETUP,0x07);           // 数据传输率 1Mbit/s,发射功率 0dBm,低噪声放大器增益
    NRFWriteReg(W_REGISTER+CONFIG,0x0f);             // CRC 使能，16 位 CRC 校验，通电，接收模式
    CE = 1;
    NRFDelay(5);
}
/********************** 检测是否有接收到数据 *****************************/
void CheckACK()
{   //用于发送模式接收应答信号
    sta=NRFReadReg(R_REGISTER+STATUS);               // 返回状态寄存器
    if(TX_DS)
        NRFWriteReg(W_REGISTER+STATUS,0xff);         // 清除 TX_DS 或 MAX_RT 中断标志
}
/********************* 接收数据 *******************************************/
void GetDate()
{
    int i;
    sta=NRFReadReg(R_REGISTER+STATUS);               // 发送数据后读取状态寄存器
    if(RX_DR)                                         // 判断是否接收到数据
    {
    CE=0;//待机
    NRFReadRxDate(R_RX_PAYLOAD,Date,RX_DATA_WITDH);// 从 RXFIFO 读取数据接收 4 位即可，后一位为结束位
    NRFWriteReg(W_REGISTER+STATUS,0xff); // 接收到数据后 RX_DR,TX_DS,MAX_PT 都置 1，通过写 1 来清除中断标志
    CSN=0;
    NRFSPI(FLUSH_RX);//用于清空 FIFO，非常关键，否则会出现意想不到的后果
    CSN=1;
    }
    //NRFWriteReg(W_REGISTER+STATUS,0xff); // 接收到数据后 RX_DR,TX_DS,MAX_PT 都置高为 1，通过写 1 来清除中断标志
    for(i=0;i<100;i++)
        Shumaguan(Date);
}

#include "reg52.h"
#include "nrf24l01.h"
#include "delay.h"
/***************** 延时函数 ************************/
void NRFDelay(unint t)
{
    unint x,y;
```

```c
    for(x=t;x>0;x--)
     for(y=110;y>0;y--);
}
```

25.4 技术拓展

NRF24L01 一对多无线发射主机实验程序如下。

```c
/*********************BST-M51 实验开发板例程 **********************
*   平台: BST-M51 + Keil U4 + STC89C51
*   名称: NRF24L01 无线送射（一对多）
*   晶振: 11.059 2MHz
***************************************************************/
#include<reg52.h>
#include "NRF24L01.h"
#include "Delay.h"

sbit CE=P1^0;      //RX/TX 模式选择端
sbit SCLK=P1^1;    //SPI 时钟端
sbit MISO=P1^2;    //SPI 主机输入从机输出端
sbit CSN=P1^3;     //SPI 片选端 // 就是 SS
sbit MOSI=P1^4;    //SPI 主机输出从机输入端
sbit IRQ=P1^5;     // 可屏蔽中断端

idata unsigned char  RxData_Buf0[RX_DATA_WITDH];      // 存储通道 0 接收到的数据
idata unsigned char  RxData_Buf1[RX_DATA_WITDH];      // 存储通道 1 接收到的数据
//idata uchar RxData_Buf2[RX_DATA_WITDH];              // 存储通道 2 接收到的数据
//idata uchar RxData_Buf3[RX_DATA_WITDH];              // 存储通道 3 接收到的数据
//idata uchar RxData_Buf4[RX_DATA_WITDH];              // 存储通道 4 接收到的数据
//idata uchar RxData_Buf5[RX_DATA_WITDH];              // 存储通道 5 接收到的数据

unsigned char  code Rx_Addr0[RX_ADDR_WITDH]={0xc0,0x43,0x10,0x10,0x01};    // 主机通道 0 的接收地址与从机 0 的发送地址相同（发送时当做发送地址，发送数据给从机 0）
unsigned char  code Rx_Addr1[RX_ADDR_WITDH]={0xc1,0xc2,0xc2,0xc2,0xc2};
unsigned char  code Rx_Addr2[RX_ADDR_WITDH]={0xc2,0xc2,0xc2,0xc2,0xc2};
unsigned char  code Rx_Addr3[RX_ADDR_WITDH]={0xc3,0xc2,0xc2,0xc2,0xc3};
//uchar code Rx_Addr4[1] = {0xc4};
//uchar code Rx_Addr5[1] = {0xc5};

unsigned char code Tx_Addr0[TX_ADDR_WITDH]={0xc0,0x43,0x10,0x10,0x01};// 发送地址
unsigned char code Tx_Addr1[TX_ADDR_WITDH]={0xc1,0xc2,0xc2,0xc2,0xc2};// 发送地址
unsigned char code Tx_Addr2[TX_ADDR_WITDH]={0xc2,0xc2,0xc2,0xc2,0xc2};// 发送地址
unsigned char code Tx_Addr3[TX_ADDR_WITDH]={0xc3,0xc2,0xc2,0xc2,0xc2};// 发送地址
```

此项目附操作视频及代码资料。

Chapter 26

项目26
智能风扇系统（综合实验）

项目目标

- 综合实验，过程中用到矩阵键盘模块、液晶屏、人体红外传感器和扇叶等，由之前所学的各部分内容组合而成，编程则用到定时器中断等内容。

建议学时

- 8学时。

知识要点

- 本项目为综合实训项目，涉及本书全部章节。

技能掌握

- 熟练使用本书配套的单片机，以及掌握各模块的程序编写以及原理。

26.1 项目分析

通过综合实训项目的实验，使学生获得更多的直接经验，综合实训项目的实验内容密切联系现实生活，注重对知识技能的综合运用，体现经验和生活对学生发展价值的实践性课程。

本项目主要通过编程实验来整合全书所学内容。

26.2 技术准备

本项目以综合全书内容的编程实验为主，要求全面掌握本书之前各项目内容。

26.3 项目实施

智能风扇实验程序如下。

```c
1118b20.c 文件
/*******************BST-M51 实验开发板例程 **********************
*   平台: BST-M51 + Keil U4 + STC89C51
*   名称: 按键扫描
*   晶振: 11.059 2MHz
***************************************************************/
#include<reg51.h>
#include<allhead.h>

sbit ds = P2^2;

void dsInit()
{

    unsigned int i;
    ds = 0;
    i = 100;
     while (i>0) i--;
    ds = 1;
    i = 4;
     while (i>0) i--;
}

void writeByte(unsigned char dat)
{
    unsigned int i;
    unsigned char j;
    bit b;
    for (j = 0; j < 8; j++)
    {
        b = dat & 0x01;
        dat >>= 1;

        if (b)
        {
            ds = 0;            i++; i++;
```

```c
                ds = 1;
                i = 8; while (i>0) i--;
            }
            else
            {
                ds = 0;
                i = 8; while (i>0) i--;
                ds = 1;
                i++; i++;
            }
    }
}

bit readBit()
{
    unsigned int i;
    bit b;
    ds = 0;
    i++;
    ds = 1;
    i++; i++;
    b = ds;
    i = 8;
    while (i>0) i--;
    return b;
}

unsigned char readByte()
{
    unsigned int i;
    unsigned char j, dat;
    dat = 0;
    for (i=0; i<8; i++)
    {
        j = readBit();

        dat = (j << 7) | (dat >> 1);
    }
    return dat;
}

void sendChangeCmd()
{
    dsInit();
    delayMs(1);
    writeByte(0xcc);
    writeByte(0x44);
}

void sendReadCmd()
{
    dsInit();
    delayMs(1);
    writeByte(0xcc);
    writeByte(0xbe);
}
```

```c
int getTmpValue()
{
    unsigned int tmpvalue;
    int value;
    float t;
    unsigned char low, high;
    sendReadCmd();

    low = readByte();
    high = readByte();

    tmpvalue = high;
    tmpvalue <<= 8;
    tmpvalue |= low;
    value = tmpvalue;

    t = value * 0.0625;

    value=t*10+0.5;
    return value;
}
//adda.c 文件
/******************BST-V51 实验开发板例程 **********************
*   平台:BST-M51 + Keil U4 + STC89C51
*   名称:LCD1602模块实验
*   晶振:11.059 2MHz
***************************************************************/
#include<reg51.h>       //包含单片机寄存器的头文件
#include <intrins.h>
#include<allhead.h>

#define   AddWr 0x90     //PCF8591 地址

// 变量定义
unsigned char AD_CHANNEL=0;
unsigned char  D[32];

sbit scl=P2^0;          //I²C 时钟
sbit sda=P2^1;          //I²C 数据
bit ack;                /*应答标志位*/


unsigned char date;


/***************************************************************
启动总线函数
函数原型: void  Start_I2c();
功能:启动 I²C 总线,即发送 I²C 起始条件
***************************************************************/
void Start_I2c()
{
 sda=1;          /*发送起始条件的数据信号*/
 _nop_();
 scl=1;
 _nop_();        /*起始条件建立时间 >4.7μs,延时*/
 _nop_();
```

```c
    _nop_();
    _nop_();
    _nop_();
    sda=0;           /*发送起始信号*/
    _nop_();         /*起始条件锁定时间>4μs*/
    _nop_();
    _nop_();
    _nop_();
    _nop_();
    scl=0;           /*钳住I²C总线，准备发送或接收数据 */
    _nop_();
    _nop_();
}

/***************************************************************
结束总线函数
函数原型：void  Stop_I2c();
功能：结束I²C总线，即发送I²C结束条件
****************************************************************/
void Stop_I2c()
{
    sda=0;       /*发送结束条件的数据信号*/
    _nop_();     /*发送结束条件的时钟信号*/
    scl=1;       /*结束条件建立时间>4μs*/
    _nop_();
    _nop_();
    _nop_();
    _nop_();
    sda=1;       /*发送I²C总线结束信号*/
    _nop_();
    _nop_();
    _nop_();
    _nop_();
}


/***************************************************************
字节数据发送函数
函数原型：void  I2C_SendByte（UCHAR c）;
功能：将数据c发送出去，可以是地址，也可以是数据，发送完毕后等待应答，并对此状态位进行操作，ack=1,发送数据正常；ack=0,被控器件无应答或损坏（即不应答，或非应答，都使ack=0)
****************************************************************/
void  I2C_SendByte(unsigned char  c)
{
    unsigned char  i;

    for(i=0;i<8;i++) /*要传送的数据长度为8位*/
    {
        if((c<<i)&0x80) sda=1;    /*判断发送位*/
        else  sda=0;
        _nop_();
        scl=1;                    /*置时钟线为高电平，通知被控器开始接收数据位*/
        _nop_();
        _nop_();                  /*保证时钟高电平周期>4μs*/
        _nop_();
        _nop_();
```

```c
        _nop_();
        scl=0;
    }
    
    _nop_();
    _nop_();
    sda=1;                    /*8 位发送完后释放数据线,准备接收应答位*/
    _nop_();
    _nop_();
    scl=1;
    _nop_();
    _nop_();
    _nop_();
    if(sda==1) ack=0;
      else ack=1;             /*判断是否接收到应答信号*/
    scl=0;
    _nop_();
    _nop_();
}


/***********************************************************
字节数据接收函数
函数原型: UCHAR  I2C_RcvByte();
功能:用来接收从器件传来的数据,并判断总线错误(不发应答信号),发完后请用应答函数应答从机。
***********************************************************/
unsigned char  I2C_RcvByte()
{
    unsigned char  retc=0,i;
    sda=1;                    /*置数据线为输入方式*/
    for(i=0;i<8;i++)
    {
        _nop_();
        scl=0;                /*置时钟线为低电平,准备接收数据位*/
        _nop_();
        _nop_();              /*时钟低电平周期>4.7μs*/
        _nop_();
        _nop_();
        _nop_();
        scl=1;                /*置时钟线为高电平,使数据线上数据有效*/
        _nop_();
        _nop_();
        retc=retc<<1;
        if(sda==1) retc=retc+1; /*读数据位,接收的数据位放入 retc 中*/
        _nop_();
        _nop_();
    }
    scl=0;
    _nop_();
    _nop_();
    return(retc);
}


/***********************************************************
应答子函数
函数原型: void Ack_I2c(bit a);
功能:主控器进行信号应答(可以是应答或非应答,由位参数 a 决定)
```

```c
*********************************************************/
void Ack_I2c(bit a)
{
    if(a==0) sda=0;                         /* 在此发出应答或非应答信号 */
    else sda=1;                             /*0 为发出应答,1 为非应答信号 */
    _nop_();
    _nop_();
    _nop_();
    scl=1;
    _nop_();
    _nop_();                                /*时钟低电平周期>4μs*/
    _nop_();
    _nop_();
    _nop_();
    scl=0;                                  /*清时钟线,钳住 I²C 总线以便继续接收 */
    _nop_();
    _nop_();
}

/*********************************************************
* 函数名:Pcf8591_DaConversion
* 函数功能:PCF8591 的输出端输出模拟量
* 输入:addr(器件地址)、channel(转换通道)、value(转换的数值)
* 输出 : 无
****************** *********************************************/
bit Pcf8591_DaConversion(unsigned char addr,unsigned char channel, unsigned char Val)
{
    Start_I2c();                    // 启动总线
    I2C_SendByte(addr);             // 发送器件地址
    if(ack==0) return(0);
    I2C_SendByte(0x40|channel);     // 发送控制字节
    if(ack==0) return(0);
    I2C_SendByte(Val);              // 发送 DAC 的数值
    if(ack==0) return(0);
    Stop_I2c();                     // 结束总线
    return(1);
}

/*********************************************************
* 函数名:Pcf8591_SendByte
* 函数功能:写入 1 个控制命令
* 输入:addr(器件地址)、channel(转换通道)
* 输出:无
*********************************************************/
bit PCF8591_SendByte(unsigned char channel)
{
    Start_I2c();                    // 启动总线
    I2C_SendByte(AddWr);            // 发送器件地址
    if(ack==0) return(0);
    I2C_SendByte(0x40|channel);     // 发送控制字节
    if(ack==0) return(0);
    Stop_I2c();                     // 结束总线
    return(1);
}

/*********************************************************
```

```c
* 函数名: PCF8591_RcvByte
* 函数功能: 读取一个转换值
* 输入:
* 输出: dat
***********************************************/
unsigned char PCF8591_RcvByte()
{
unsigned char dat;

    Start_I2c();                // 启动总线
    I2C_SendByte(AddWr+1);      // 发送器件地址
if(ack==0)return(0);
    dat=I2C_RcvByte();          // 读取数据0

    Ack_I2c(1);                 // 发送非应答信号
    Stop_I2c();                 // 结束总线
return(dat);
}
/*-----------------------------------------------
把读取值转换成一个一个的字符,送串口显示
-----------------------------------------------*/
void To_ascii(unsigned char num)
{
    SBUF=num/100+'0';
    delayMs(200);
    SBUF=num/10%10+'0';
    delayMs(200);
    SBUF=num%10+'0';
    delayMs(200);
}
//delay.c
/*******************BST-M51 实验开发板例程 **********************
* 平台: BST-M51 + Keil U4 + STC89C52RC
* 名称: 按键扫描
* 晶振: 11.059 2MHz
***********************************************************/
#include <intrins.H>
void delayMs(unsigned int xms)
{
  unsigned int i,j;
  for(i=xms;i>0;i--)          //i=xms 即延时 xms
    for(j=112;j>0;j--);
}

void delayUs()
{
    _nop_();
}
// keyscan.c
/*******************BST-M51 实验开发板例程 **********************
* 平台: BST-M51 + Keil U4 + STC89C52RC
* 名称: 按键扫描
* 晶振: 11.059 2MHz
***********************************************************/
#include<reg51.h>
#include<allhead.h>
```

```c
unsigned char keyscan()
{
    static unsigned char temp,key;

    //////////////// 第 4 列扫描 ////////////////////
    P3 &= 0xfe;//1111 1110 让 P3.0 端口输出低电平
    temp=P3;
    temp=temp&0xf0;//1111 0000 位"与"操作屏蔽后 4 位
    if(temp!=0xf0)
    {
        delayMs(10);//消抖
        temp=P3;
        temp=temp&0xf0;
        if(temp!=0xf0)
        {
            temp=P3;
            switch(temp)
            {
                case 0x7e:                      //1110 1110 "A"被按下
                        key=level1;
                        break;
                case 0xbe:                      //1101 1110 "B"被按下
                        key=level2;
                        break;
                case 0xde:                      //1011 1110 "C"被按下
                        key=level3;
                        break;
                case 0xee:                      //0111 1110 "D"被按下
                        key=turnoff;
                        break;
            }
            while(temp!=0xf0)
            {
                    temp=P3;
                    temp=temp&0xf0;
            }
            return key;
        }
    }
    else
    return nokey;
}


// lcd1602.c
/********************BST-M51 实验开发板例程 *********************
*   平台: BST-M51 + Keil U4 + STC89C52RC
*   名称: 按键扫描
*   晶振: 11.059 2MHz
************************************************************/
#include<reg51.h>
#include<allhead.h>
#include <math.H>

sbit rw = P1^1;
```

```c
sbit RS = P1^0;
sbit LCDEN = P2^5;

void writeComm (unsigned char comm)
{
    RS = 0;
    P0 = comm;
    LCDEN = 1;
    delayUs();
    LCDEN = 0;
    delayMs(1);
}

void writeData (unsigned char dat)
{
    RS = 1;
    P0 = dat;
    LCDEN = 1;
    delayUs();
    LCDEN = 0;
    delayMs(1);
}

void writeString (unsigned char * str,unsigned char length)
{
    unsigned char i;
    for (i = 0; i < length; i++)
    {
        writeData (str[i]);
    }
}

void lcd_init()
{
    rw = 0;
    writeComm (0x38);
    writeComm (0x0c);
    writeComm (0x06);
    writeComm (0x01);
}

void display (unsigned char speed,int v)
{
    unsigned char datas[9];
    unsigned char level[7];

    level[0]= '1';
    level[1]= 'e';
    level[2]= 'v';
    level[3]= 'e';
    level[4]= '1';
    level[5]= ' ';
    level[6]= '0' +speed;
    writeComm (0x80+4);
    writeString (level, 7);
```

```c
    datas[0] = 'T';
    datas[1] = 'E';
    datas[2] = 'M';
    datas[3] = 'P';
    datas[4] = ':';
      datas[5] = '0' + (v/100)%10;
      datas[6] = '0' + (v/10)%10;
      datas[7] = '.';
      datas[8] = '0' +v%10;
writeComm(0xc0+3);
writeString(datas, 9);
}


// recive.c
/*******************BST-V51 实验开发板例程 ********************
* 平台：BST-M51 + Keil U4 + STC89C52RC
* 名称：红外接收模块实验
* 晶振：11.059 2MHz
* 说明：免费开源，不提供源代码分析
*************************************************************/
#include<reg51.h>
#include<allhead.h>

sbit IR=P3^2;    // 红外接口标志
unsigned char  irtime;// 红外用全局变量


bit irpro_ok,irok;
unsigned char IRcord[4];
unsigned char irdata[33];


//main.c
/*******************BST-M51 实验开发板例程 ********************
* 平台：BST-M51 + Keil U4 + STC89C52RC
* 名称：智能风扇
* 晶振：11.059 2MHz
*************************************************************/
#include<reg51.h>
#include<allhead.h>

#define Coil_CD1 {A1=1;B1=1;C1=0;D1=0;}//CD 相通电，其他相断电
#define Coil_AD1 {A1=0;B1=1;C1=1;D1=0;}//AD 相通电，其他相断电
#define Coil_AB1 {A1=0;B1=0;C1=1;D1=1;}//AB 相通电，其他相断电
#define Coil_BC1 {A1=1;B1=0;C1=0;D1=1;}//BC 相通电，其他相断电
#define Coil_A1  {A1=0;B1=1;C1=1;D1=1;}//A 相通电，其他相断电
#define Coil_B1  {A1=1;B1=0;C1=1;D1=1;}//B 相通电，其他相断电
#define Coil_C1  {A1=1;B1=1;C1=0;D1=1;}//C 相通电，其他相断电
#define Coil_D1  {A1=1;B1=1;C1=1;D1=0;}//D 相通电，其他相断电
#define Coil_OFF {A1=1;B1=1;C1=1;D1=1;}// 全部断电


sbit A1=P1^4;  // 定义步进电机连接端口
sbit B1=P1^5;
sbit C1=P1^6;
sbit D1=P1^7;
sbit IR=P3^2;    // 红外端口标志
```

```c
unsigned char speed=0,key0=0,adda,key,irtime,irdata[33],IRcord[4];
int tmp;
unsigned int count=0;
bit irpro_ok,irok;
/*-----------------------------------------------
红外码值处理
-----------------------------------------------*/
void Ircordpro(void) // 红外码值处理函数
{
  unsigned char i, j, k;
  unsigned char cord,value;

  k=1;
  for(i=0;i<4;i++)           // 处理 4 字节
    {
     for(j=1;j<=8;j++)// 处理 1 字节 8 位
        {
         cord=irdata[k];
         if(cord>7)// 大于某值为 1,这个和晶振有绝对关系,这里使用 12MHz 计算,此值可以有一定误差
            value|=0x80;
         if(j<8)
           {
            value>>=1;
           }
          k++;
         }
      IRcord[i]=value;
      value=0;
     }
    irpro_ok=1;// 处理完毕标志位置 1
}
/*-----------------------------------------------
键值处理
-----------------------------------------------*/
unsigned char Ir_work(void) // 红外键值散转程序
{
  unsigned char key;
  switch(IRcord[2]) // 判断第 3 个数码值
   {
     case 0x0c:key=level1;break;//1  显示相应的按键值
     case 0x18:key=level2;break;//2
     case 0x5e:key=level3;break;//3
        default:break;
   }
  irpro_ok=0;// 处理完成标志
  return key;
}
/*-----------------------------------------------
定时器 T0 初始化子程序
-----------------------------------------------*/
void timer0_init(void)
{
  TMOD |= 0x01;    // 定时器 T0 工作方式 2,TH0 是重装值,TL0 是初值,使用"|"符号可以在使用多个定时器时不受影响
  TH0=0xFC;               // 给定初值 1ms
  TL0=0x66;
  EA=1;                     // 总中断打开
```

```c
    ET0=1;              // 定时器中断打开
    TR0=1;              // 定时器开关打开
}
/*------------------------------------------------
定时器 T1 初始化子程序
------------------------------------------------*/
void timer1_init (void)
{
    TMOD |= 0x01;       // 使用模式 1, 16 位定时器, 使用"|"符号可以在使用多个定时器时不受影响
    TH1=0xFC;           // 给定初值 1ms
    TL1=0x66;
    EA=1;               // 总中断打开
    ET1=1;              // 定时器中断打开
    TR1=1;              // 定时器开关打开
}
/*------------------------------------------------
外部中断 0 初始化
------------------------------------------------*/
void ex0init (void)
{
    IT0 = 1;            // 指定外部中断 0 下降沿触发, INT0 (P3.2)
    EX0 = 1;            // 使能外部中断
    EA = 1;             // 开总中断
}
/*------------------------------------------------
主函数
------------------------------------------------*/
void main()
{
    timer0_init();
    timer1_init();
    Coil_OFF
    lcd_init();
    sendChangeCmd();
    while(1)
    {
        key0 = keyscan();

        if(key0 != nokey)
        {
            speed = key0;
            key = key0;
        }
        PCF8591_SendByte(2);
        adda=PCF8591_RcvByte();
        if((adda>150) && ((key0 == turnoff)||(key0 == nokey)))
        {
            if(tmp>340)
                speed=level3;//t=5;
            else if(tmp>320)
                speed=level2; //t=6;
            else if(tmp>300)
            speed=level1;//t=7;
        }
        else if((!key) && (!adda))
            speed=turnoff;
```

```c
    }
}
/*-----------------------------------------------
定时器 T1 中断子程序
-----------------------------------------------*/
void Timer1_isr(void) interrupt 3
{
  static unsigned char times,i;
  unsigned char t;
  TH1=0xFC;                  // 重新赋值 1ms
  TL1=0x66;

  switch(speed)
   {
    case level1:t=40;break;
    case level2:t=30;break;
    case level3:t=20;break;
    default:break;
   }
  if((speed)||(adda>150))
   {
    if(times==t)
     {
       times=0;
       switch(i)
        {
         case 0:Coil_A1;i++;break;
         case 1:Coil_B1;i++;break;
         case 2:Coil_C1;i++;break;
         case 3:Coil_D1;i++;break;
         case 4:i=0;break;
         default:break;
        }
     }
   times++;
   }
}
/*-----------------------------------------------
定时器 T0 中断子程序
-----------------------------------------------*/
void Timer0_isr(void) interrupt 1
{
  static unsigned char tt;
  TH1=0xFC;                  // 重新赋值 1ms
  TL1=0x66;
  if(tt==5)
   {
   tmp=getTmpValue();
   display(speed,tmp);
   sendChangeCmd();
   tt=0;
   }
  tt++;
}
```

```c
/*-----------------------------------------------
外部中断 0 中断处理
-----------------------------------------------*/
void EX0_ISR (void) interrupt 0          // 外部中断 0 服务函数
{
  static unsigned char  i;               // 接收红外信号处理
  static bit startflag;                  // 是否开始处理标志位

  if (startflag)
  {
    if (irtime<17&&irtime>=9)// 引导码 TC9012 的头码, 9ms+4.5ms
    i=0;
    irdata[i]=irtime;// 存储每个电平的持续时间, 用于以后判断是 "0", 还是 "1"
    irtime=0;
    i++;
    if (i==33)
    {
     irok=1;
     i=0;
    }
  }
  else
  {
    irtime=0;
    startflag=1;
  }

}
```

附件
亚博BST-M51主要模块电路图

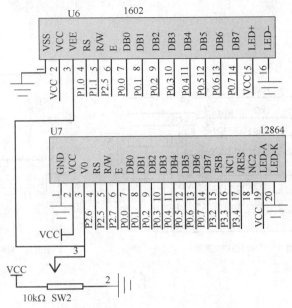

附图1 显示模块 1602/12864 模块电路图

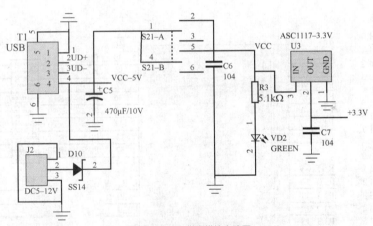

附图2 USB 供电模块电路图

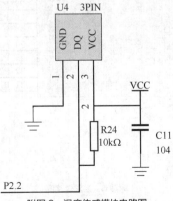

附图3 温度传感模块电路图

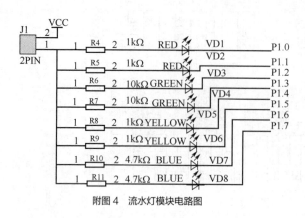

附图 4 流水灯模块电路图

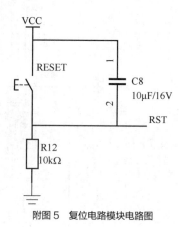

附图 5 复位电路模块电路图

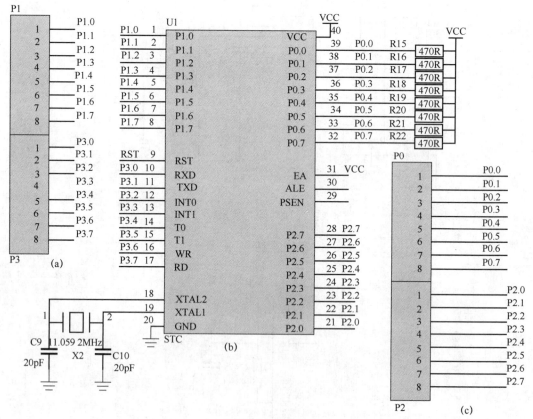

附图 6 系统电路图

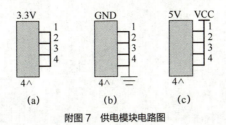

附图7　供电模块电路图

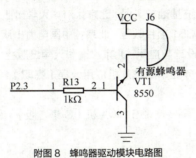

附图8　蜂鸣器驱动模块电路图

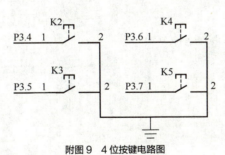

附图9　4位按键电路图

参考文献

[1] 王东锋,王会良,董冠强.单片机C语言应用100例.北京:电子工业出版社,2009.
[2] 彭伟.单片机C语言程序设计实训100例.北京:电子工业出版社,2009.
[3] 张铮.单片机与嵌入式系统基础与实训.北京:清华大学出版社,2011.
[4] 张晓峰,郭显久.单片机C51项目教程.北京:中国电力出版社,2011.
[5] 王小建,胡长胜.单片机设计与应用.北京:清华大学出版社,2011.
[6] 程院莲,廖春蓝.基于任务驱动的单片机应用教程(高职).西安:西安电子科技大学出版社,2011.
[7] 张义和,王敏男,许宏昌,等.例说51单片机(C语言版).北京:人民邮电出版社,2010.